DON'T GET RIPPED OFF

DON'T GET RIPPED OFF

by
MIKE RAWLINSON

THE COMPLETE USER-FRIENDLY GUIDE TO BUYING A SECOND-HAND CAR

Published in May, 1989 by Shaw & Sons Ltd
Shaway House, London SE26 5AE

ISBN 0 7219 1042 4

Text typeset by Goodfellow and Egan, Cambridge
Printed in Great Britain by The Gordon Press, Croydon

Contents

Preface

It began with motorcycles. I paid £4 for my first one – a large pile of assorted pieces of metal which I brought home in a wheel-barrow. At that stage all I knew was that if all these bits and pieces were assembled in precisely the correct way, the thing would somehow go. A few weeks later I was driving around, and feeling very pleased with myself.

Later it was cars, but as time progressed, the novelty of spending my evenings with muck up to my elbows wore off, so I gave them up. At least I tried to, but I have a lot of friends. Every time someone wanted to buy a car, they'd ask me to go along with them to make sure it was OK. I'd try to get out of it, but a friend in need can be very persuasive, particularly when they've softened up the defences at the bar of the local pub. I began to charge exhorbitant fees for my services, but that only seemed to encourage them more.

I tried to tell them it wasn't difficult, explaining that you don't actually need to know how it all works, just what to look for. One day someone suggested that I write a book on the subject. Sometimes the most obvious solutions to problems are the most difficult to see!

This book is dedicated to all my friends who from now on will have to carry out their own car inspections.

Thanks to Gail who painstakingly read the

manuscript, corrected all the mistakes, and stuck with me throughout, without the benefit of an instruction manual.

PART I

Introduction

Every few years we face a weighty task. For some it is a time of great excitement, whilst for others it is a time of worry and apprehension. Most of us, however, greet this great undertaking with rather mixed feelings and simply hope that it will all work out all right. No matter how we feel about it there comes a time when we can put it off no longer, and we have to face the fact that it's time to get rid of the old jalopy and buy a new car. Unfortunately for the majority of us, a new car doesn't actually mean a new car, but a replacement car: a second-hand car, or as is more frequently the case, a third- or fourth-hand car.

The reason we are not buying a brand-new car will usually be because we simply cannot afford it, but there is also a sound financial argument for buying a used vehicle. The value of a new car falls very rapidly during the first two years of its life, and it is not uncommon for a two-year-old car to be worth only half the cost of a new one. That works out at rather a lot of money just for the status of having this year's registration letter tacked onto the back of your vehicle, money that could perhaps be better used on something else.

So, for those of us who are not status conscious, there is a lot to be said for looking round for a really good cast-off vehicle. The problem with used cars is that they are offered for sale because they are no longer wanted by their owner, and there will be a reason why they are no longer wanted. Unless you are careful, you could end up buying someone else's problems.

People decide to get rid of their old car for a wide variety of reasons. Perhaps the family has grown, and the car is no longer quite big enough to fit everyone in. Or perhaps redundancy has necessitated a slight drop in living standards, and the second car has had to go. Some people change their car every year or two as a matter of course, without considering the reason why they do it. Others may have had a salary increase, and decide that they must now be seen in a rather more prestigious vehicle. There are certainly plenty of reasons why people sell cars, but one of the more popular ones has not yet been mentioned.

There is, rather sadly, a prevailing attitude amongst the motoring public that says that when something looks like going wrong with a car and it looks like it will cost a lot to put matters right, then you have to choose one of two alternatives. Either you get it fixed and pay a large bill, or sell it to someone else. Provided that the problem isn't too obvious, or can be disguised then many opt for the second alternative.

Some car owners are honest, and will tell you if there is something wrong with the vehicle they are selling. There is, however, something about motor cars which brings out the worst in people. For some it comes out in the way they drive: a normally placid and quiet person can become aggressive and belligerent as soon as they get behind the wheel of a car. In others the fact that they are dealing with a motor vehicle will cause them to suspend their normal standards of honesty. Even the most upright citizen will sometimes forget to mention one or two problems that he has been having with his car, if he thinks someone is going to buy it from him.

On top of this, there is the problem of dishonest car dealers. Although there are many reputable and reliable motor traders, the second-hand motor trade seems to have attracted a fair number of sharks and money-grabbers. We read terrible tales in newspapers of vehicles being sold complete with 12 months MOT, which are falling apart and not fit to be on the public highway. Some cars, fit only for the scrap heap, are resurrected by back-street workshops, bodged up, repainted and offered for sale at what are often rather fancy prices.

Then there are those stories we hear of how cars are built up from two or three vehicles which have been written off in serious accidents. Various undamaged parts are assembled together, welded up and then repainted, to make one vehicle: a vehicle which may be twisted and highly dangerous.

We hear too of those cars which, although only two or three years old, have covered very high mileages. You've seen them in your rear view mirror on the motorway: flashing their headlamps at you to pull over, then tearing past at 100 m.p.h. These cars are often owned by companies and are driven up and down the motorways, day after day, until they are traded in and replaced with a new model. There is of course nothing wrong with that: cars are meant to be used. However, these vehicles will often end up in motor auctions and may be bought up by shady dealers who are not averse to altering the milometers and selling them off as low mileage bargains to an unwary buyer. These cars will look fine and may still look pretty new from the outside, but the mechanical parts will be rather worn and whoever ends up with a car like this will spend a lot of money on repair work.

How do these unscrupulous dealers get away with it?

The answer to this is very simple. Most people, despite the fact that they spend a large part of their lives driving round in one, know little or nothing about the workings of a motorcar. They may be able to carry out a few routine tasks such as checking the tyre pressures and an ambitious owner may even manage to change the oil periodically, but as to how the thing actually works most of us haven't a clue.

Since they don't know how a car works, most people assume that a full and thorough examination of a used car is beyond their abilities and will instead put their trust in the integrity of the seller, rely on luck and good fortune, or get someone else to check it over for them. It is of course quite reasonable to assume that in order to examine a car properly, you need to know how it all works, but this is not necessarily the case. It would, after all, be just as reasonable to presume that some knowledge of the innermost secrets of the internal combustion engine are a necessary pre-requisite for the driving of a car, were it not for the fact that we already know that millions of people drive around without knowing the first thing about how the vehicle they are driving actually works. In order to learn to drive, all you need is someone to tell you what to do, and when to do it. The same applies to examining a car. Provided that you are clearly told what to do, it doesn't matter at all that you don't fully comprehend what is going on.

That's what this book is about. Using non-technical language which can be easily understood, it will take you systematically through a step-by-step procedure which is designed to pick up almost any fault in just about any second-hand car.

It doesn't matter who you are. Even if you don't know a piston ring from a track rod end, or you think that a differential is only something to do with wage negotiations, you can still do it. Provided you can drive, and can perform simple tasks akin to the boiling of an egg, or changing a vacuum cleaner dust bag, then you are quite capable of examining a motor vehicle.

You will have to be prepared to get your hands dirty (very dirty!), and you'll need to use your eyes and ears. However, if you follow the procedures given here, you will quickly determine whether a vehicle is any good or not and need never buy a duff car again.

For the benefit of those readers who do have some knowledge of the mysteries of the motorcar, some explanations are given, which do use a modicum of technical jargon. These digressions are infrequent, and the jargon is kept to an absolute minimum. Don't worry if you do not understand these sections: they are for interest only.

As you are doubtless aware, there are motor engineers who will, for a fee, examine a car for you. It must in fairness be said that with their knowledge and experience, they are likely to be in a better position to assess the condition of a vehicle than someone with no experience at all. The trouble is, they have to make a living like anyone else, so it will cost quite a lot (plus VAT) to have a car professionally inspected. Added to this, if you do wish to employ an inspector, you will have to arrange an appointment. And by the time he comes to see it, the car may well have been sold to someone else, particularly if it's a good buy.

The other problem with such inspections is that with so many duff cars around, you may end up paying for several examinations before you find a car which is worth buying. You could end up spending hundreds of pounds having useless cars examined, money which could have been spent on buying a decent car! By carrying out your own thorough inspection, you will eliminate all the defective vehicles, and if you do decide that you want a professional examination, you will probably only have to pay one inspection fee.

Finding a suitable car

Before you can contemplate spending your hard-earned overdraft on a car, you have to find one, and before you can do that, you have to decide what you want.

It's always difficult to decide what type of car to go for and you would be well advised not to have too much of a fixed idea when you begin your search. Although a brand new Ford Escort is much the same as any other new Ford Escort, a used car is very much an individual. It may once have been identical to the one in the next street, but after a couple of years of use (or misuse), the two cars will be totally different. You may have had very satisfactory service from a particular type of car in the past, but that does not mean that you could expect the same when you buy another. It all depends on how well it has been cared for. What you must look for in a car, is a vehicle in the best condition for what you can afford.

A valuable aid to deciding on which car is best for you is published by The Consumers Association entitled, *Which Car? Buying Guide* which gives detailed information on all popular cars currently available. Of particular interest is the section on used cars which gives information about the bad points for each model, so you will know what to look for when you go to examine one. You will probably be able to read a copy in your local reference library, but it is worthwhile obtaining a copy.

Another publication in which you could invest is one of the price guides available from newsagents. These will tell you what a particular car is likely to be worth, known as the 'book value'. However, these guides are national publications, and there is considerable regional variation in the price of used cars. Some people actually make a living by exploiting this fact, buying in one area, and selling elsewhere. Nevertheless, a price guide is useful, and you should avoid paying more than the 'book price' if possible.

In deciding which type of car to go for there are many factors to be considered. The first will of course be how much you can afford; but this may not be as straightforward as it seems because you will also have to spend a lot on running the car once you have bought it. A car which is cheap to buy may in the long run cost more if the running costs are high, so what you need to try to work out is your overall annual motoring costs, including the cost of buying and replacing your car. There are several factors which must be considered before deciding on the type of car you want.

■ Insurance
Do find out what this is likely to cost before you decide to take the plunge. You may rather fancy going round in a flashy GTX Turbo XR 14 model, but the insurance costs can be phenomenal, particularly for young drivers. It's a good idea to phone an insurance broker to obtain a quotation for the type of car you have in mind. Generally the higher the performance of the car, the higher the insurance costs, but there are other factors: foreign cars are often more expensive due to the relatively high cost of accident repairs and some companies offer a discount on older cars.

■ Fuel consumption
Obviously you will want to spend as little as possible on petrol, but if your annual mileage is fairly low, an economy class car may not necessarily be the best choice. Such cars have a high value simply because of their low fuel consumption and you can spend rather a lot of money on something which is getting on a bit and may require a fair bit of repair work before long. Cars which guzzle a lot can be relatively cheap to buy, particularly if they are getting on in years and if you only intend to use it

for local shopping trips it could in the long run be cheaper to get a bigger car, although of course, insurance will be dearer.

I have a friend who bought an ageing low-mileage Rover 3.5 litre for about the same as he would have expected to pay for an average Mini. He only needed third party insurance, and although it didn't go very far on a tankful, he did have a few years luxury motoring at a cost comparable to going round in a 'tin-can-on-wheels' economy car.

It is worth mentioning here that the motorcar is responsible for many environmental problems which we are now having to face up to, not the least of which is that of lead pollution. Unlike its predecessor the horse, which produced large quantities of fertilizer as a by-product, the motorcar by contrast is slowly poisoning the air we breathe and the land on which we grow our food. It is now well known that airborne lead from car exhausts affects the development of young children, and we must surely do all we can to reduce this problem. The government has responded by reducing the tax on lead-free petrol and has recently followed the lead taken by other European Countries in making lead-free fuel considerably cheaper.

So it makes a lot of sense to buy a car which will run on lead-free petrol, or one which can easily be converted. For some cars conversion is simple and costs very little. Others, including sad to say, many British cars, are impossible to convert. If you want to find out whether a car you are interested in can be run on lead-free, then a quick phone call to the local main dealer for that make should give you the information you need about the costs, if any, of conversion.

■ Repairs
Some cars cost more to repair than others. There are some models which seem to go wrong fairly regularly, whilst others seem to go on and on with very little attention. Spare parts for foreign cars generally cost more and in some cases can be difficult to obtain. The *Which Car? Buying Guide* is again useful as it gives information about the reliability of most cars, along with details of spare parts availability. In addition it tells you which cars are easy to repair yourself and which are not. If you are a keen do-it-yourselfer, this is worth taking account of as labour charges in garages can be extremely expensive.

If you can do your own repairs, it is quite acceptable to save money by obtaining second-hand parts from a scrapyard, although it would be unwise in the case of brakes, steering and other safety systems to use parts which are not new. If you can fix your own car then it is worth buying a popular model to ensure a plentiful supply of cheap replacement parts. Obviously the main reason that older cars cost less is that they need to be repaired more often. If you have to rely on garages to do this for you, you may in the long run end up spending more than if you had bought a better car in the first place.

■ Colour
This may not seem to be very important, but it is well known among car dealers that red cars are much easier to sell, closely followed by white, whilst green or brown cars can be very difficult to get rid of. If you don't intend to keep the car long then go for a popular colour. If on the other hand you'll be together till the bitter end, then you may be in a position to haggle successfully over the

price of a car with a less than desirable colour scheme. I once bought a car from a dealer which was the most hideous shade of yellowy-green. I paid a lot less than book value and it went well for many years. Avoid metallic colours as they are difficult to match if repair work needs to be done. Black cars show every minor scratch and blemish and are difficult to repair without it showing. Scratches and repairs show least on white cars, which means you'll have to be extra careful when you look at one.

■ Depreciation
This is the loss of value as the car gets older. Not much you can do about this, but some cars hold value better than others. This means that for cars with low depreciation you'll get a reasonable price when you sell, but you'll have to pay more to buy one. Cars made in Eastern Europe can depreciate very quickly. Again the *Which Car? Buying Guide* gives details.

■ Your Needs
You must of course think carefully about the sort of car you will need. Only you can decide what your requirements are in terms of size, performance, etc, but do beware of buying anything with fancy gimmics like electrically operated windows which you don't really need. Not only will you have to pay more, but there is potentially more to go wrong.

Sources of used cars

There are three main sources of used motor vehicles: car dealers, private sales and motor auctions. Whoever you buy from (auctions can be an exception), it is an offence in law for them to sell a vehicle in an unroadworthy or dangerous condition, unless it is specifically sold for spare parts.

This of course does not prevent people from selling unroadworthy vehicles, but it does mean you have some legal protection against whoever you buy from.

Dealers

If you buy from a motor trader, you'll probably have to pay more, but you will in theory have some comeback if the vehicle turns out not to be quite the epitome of mechanical perfection it was cracked up to be. It has to live up to the description it was given by the seller and has to be suitable for the purpose for which it was sold.

You may be offered a warranty with a used car. This is a type of guarantee which will cover certain mechanical faults for a certain period. It is, in effect, an insurancy policy which the dealer has to pay for. In some instances you may be able to buy a warranty yourself at the time of sale. Do read the small print very carefully. It is not uncommon for a warranty to exclude any faults present at the time of purchase which a motor engineer would be expected to find. Also excluded will be any faults arising through normal wear, lack of maintenance, misuse etc. Most electrical equipment, the bodywork and many ancilliary components are also likely to be excluded. In other words, some warranties are hardly worth having, so don't assume that because a car is sold with a warranty that you don't need to make a thorough inspection. Both vehicle and the warranty should be scrutinised with great care.

Beware too of a car which is being offered for sale privately which is 'still under warranty'. It may well be that someone is selling a car on which the warranty has not yet expired, but you may find buried in the small print a clause which says that the warranty is 'not transferable', or that it 'cannot be assigned'. This simply means that if the car is sold, the warranty expires and the new owner cannot claim for any repairs under it.

If the vehicle is sold without a warranty, then you still have legal rights if the car turns out to be somewhat inferior to the way the seller described it. The problem here is that it is all very well having rights, but if you bought a duff car from a dealer, then he was probably a shark and he isn't going to give in too easily. You'll need to be very assertive, and may have to take legal action which is expensive, time consuming, very worrying and still leaves you with a duff car while it is all being sorted out.

If you do decide to buy from a dealer, make sure that he belongs to one of the motor trade associations, membership of which will be indicated by a certificate on the office wall.

Be prepared to haggle over the price as the selling price of a used car will be often be based on the assumption that a ropey trade-in will have to be accepted. If you are not part-exchanging then you can expect a substantial price reduction. You may also get a lower price if you pay the full cash price rather than using one of the finance deals offered, particularly a very low-interest hire purchase agreement. There is no such thing as 0% interest: it simply means that you are paying a higher price for the car. It is better to negotiate a lower price and borrow the money from your bank. Bank interest rates are always lower than you will get on a finance deal.

Private sales

Prices tend to be lower if you buy privately, but you do have fewer rights if the car you buy turns out to be a 'clapped-out' wreck. Of course, if there are any major problems with the vehicle, then you won't be buying it because you'll soon be proficient at vehicle examinations that few faults will escape your prying eyes!

Auctions

Auctions can be very dodgy. If a car dealer finds himself with a trade-in that isn't up to much, then it goes off to auction. Individuals usually know they'll get less for a good car if it goes into an auction, so they are more likely to sell it themselves unless it is really worn out. High mileage company cars end up in auctions, as do ex-hire and fleet cars. It's a fact that a lot of 'clapped-out' wrecks end up being sold at motor auctions, so great care is called for. Although you can hear the car running and you are usually able to inspect the underside prior to the sale, you cannot road test the car and it is on the road test that many faults will show up.

Sometimes you have a few hours after the purchase in which to return a car if it turns out to have faults which were not mentioned by the auctioneer. This is adequate time for a road test, so by buying from an auction, you will sometimes get a bargain, if you are prepared to put up with the hassle of returning a car which wasn't all it appeared to be when it came under the hammer.

If you do decide to buy from an auction, do make sure that you know what you are doing, and that you fully understand the conditions of sale. Listen to what the auctioneer says. If he describes the vehicle as having a dodgy gearbox, then it's no use finding out on the road test that it will only go into first gear and expecting them to take it back. If you buy it with a fault that was mentioned, then it's your car. Beware too of the many tricks which people try to pull off when selling through auctions: a car may have a badge on the back indicating that it is the super luxury model, but it may just be a standard model with a fancy badge. Similarly, if the knob on the gearstick is a 5-speed one, do ensure that the car does actually have five gears.

Remember, if you go to examine a car at one of the traders or in a private sale, and you're not too sure about it, then you should not buy it. If you decide you're not happy wth an auctioned vehicle, then it may be too late.

You may be offered a vehicle which has been part of a fleet e.g. an ex-hire car, or a van which belonged to the Post Office. Some people advise that they are a good buy because they are well maintained. Although they are well maintained, such vehicles have usually been hammered and are often 'clapped-out'. They are sold off because they are no longer economic to run, despite the fact that these organisatons usually have their own maintenance departments.

There can also be another problem, particularly with fleet vans. Because they are purchased in large batches, they are supplied to the specification of the purchaser, rather than the standard specification. This can mean that when you buy replacement parts such as oil filters or brake linings, they simply don't fit. Avoid these horrors!

The best buy

The price of a used car is basically determined by its age, its mileage, and its condition. By some strange irony, the factor which has the greatest effect on the cash value is the one which least affects the real worth of a car, and that is its age. A car which is five or six years old is usually worth only about 25%–50% of the new price, regardless of whether it has been used much, or how well it has been looked after. A six year-old car that has only been used to go to church on Sundays, has been in the garage all week, has been serviced and polished with loving care is still only worth the price of a six year-old model.

So, the best type of car to look for is without doubt one which has had only one owner from new, has a low mileage, and has been well cared for.

It is possible to buy a used car which has only travelled about a quarter of the distance it will travel in its life, for about a quarter of the cost of a new one, or to put it another way, you can buy a car with about three quarters of its life left (in mileage terms) for only a quarter of the price it would cost to buy new.

Such cars are rare, but they are well worth looking out for. They are usually owned by people who are fairly well off and are normally traded in against a new car after a few years of local running about. You are therefore more likely to find one at an up market dealer's showroom than in a private sale. You'll pay top price for it, but it will still be a bargain.

You may feel that this sort of car is outside your price range, but if you get it right and find a good one, it should last you for several years. In the long term this can often work out a lot more economical than a succession of cheaper cars, so it is worth planning accordingly if you can.

That's about it! You are almost ready to go out and buy your next car, but before you actually rush out and start looking, a small amount of preparation is called for.

Prepare yourself

Most people who go to buy a used car set off to

look without any clear idea of what they will do when they actually see the vehicle. Usually they will rely on a general impression, or even worse, will simply trust in what the person selling the car has to say about it. They may consider factors such as the quality of the radio, whether it has a sunshine roof, and if the price is low enough and of course, whether the colour is to their liking. The astute buyer on the other hand goes out well prepared for the task. You are going to examine your next vehicle with such thoroughness that all its faults will be revealed, and this requires some preparation.

It may at first seem odd that the way to go about buying a decent vehicle is to look for the faults, but that's the only way to do it. Far too many car buyers will see a lot that they like about the vehicle on offer, and become so enthusiastic that they miss all the faults. Car salesmen are very aware of this, and will always try to bring any favourable points to the attention of a prospective purchaser, in the hope that any shortcomings the vehicle has will not be noticed.

A good used car is not so much a car with a lot of good features, but quite simply a car which doesn't have anything wrong with it (not much anyway). Therefore the only way to find out whether a car is any good or not is to systematically go over it looking for faults. By a process of elimination, you will at some stage find a vehicle which, even if it is not quite perfect, has an acceptable degree of inadequacies, and you've found your next car.

Searching for faults

Before your searching gaze is unleashed onto the used car market, it is as well that you understand just a little of what it is you are looking for in your search for the hidden flaws of a second-hand motor car. Nothing too technical of course: a simple explanation will suffice.

As far as second-hand cars are concerned, there are basically only two things which go wrong and therefore only two things you will be looking for. That may sound ridiculous, there are after all thousands of things which can go wrong with a car, but they can usually be attributed to one of two factors: wear and corrosion.

The motorcar consists, as we know, of thousands of different parts, mostly made of metal, which when fitted together in a particular way will make this remarkable machine capable of transporting us around. In the process of performing this amazing feat, a large number of metal parts move about and rub against each other. It is inevitable that, as a result of this moving contact, some wear will take place, just as you wear out your shoes by walking in them.

This wearing out of the workings of a car is very much reduced by maintaining a film of oil or grease between the moving parts, so that they are lubricated. It is essential that oil is changed regularly if a vehicle is going to last, and that greasing and other maintenance has been properly carried out.

The first problem encountered with any used motorcar is that some wear will have taken place. What must be determined during the examination is whether excessive wear has occurred in any part of the vehicles mechanism, and whether it matters very much. Some bits and pieces do wear out long before the car as a whole is worn out, and these can often be replaced at reasonable expense. Other parts can be ruinously expensive to replace, and it is most important that the inspection reveals any such faults.

The other major thing that goes wrong with cars is due to the fact that the metal used to make the body of the vehicle is mild steel. Even cars with aluminium or fibreglass bodywork are built onto a structural chassis made from steel. Although this metal has the advantages of being strong, relatively cheap and durable, it has the unfortunate disadvantage of going rusty on contact with water.

All cars rust away eventually and any used car will have some rust somewhere (usually where it can't be seen). What must be determined is whether the degree of corrosion present on a vehicle is sufficient to be likely to cause problems in the foreseeable future. The amount of rust which is acceptable will to some extent depend on the price you will be paying, but you should never consider buying a vehicle which has a serious degree of rust, no matter how cheap, as such a vehicle may be dangerous and in any case will not last for very long.

There are a few other things to look out for such as ensuring that the seller owns the vehicle and hasn't stolen it. Or whether it has been patched up following a serious accident, or even if it is the vehicle it claims to be, and isn't something made up of two or more crashed or stolen cars. But your main concern is to look for signs of excess wear between moving parts, and corrosion of the body of the vehicle.

You won't uncover faults if you can't see what you are doing, so it is essential that you arrange to carry out your inspection during daylight – a winter evening probing with a torch is not good enough. It's going to take an hour or so to do the job, so make sure that you allow enough time.

Things to take

There are a few things which you'll need to take

with you (nothing fancy or expensive) 'high tech' test equipment will not be needed for this job.

The first, and by far the most important item to take is a *friend*. Going to look at a car can be very daunting, particularly if it is being sold by someone who is trying to pull the wool over your eyes and sell you a duff car. A friend can offer assistance, moral support, a second opinion and an extra pair of eyes and ears. It can also be useful to have someone to witness claims made by a salesman, in case any should turn out to be untrue.

The next item which is a must in the vehicle examination kit is a *clipboard and pen*. At the back of this manual is a set of test sheets which are to be used in conjunction with the text and copies can be filled in as the tests proceed. It is far easier to do this if they are on a clipboard. You may get just a little bit grubby doing an inspection, and groping round in your pocket for pen and paper is a little awkward. Besides, as I have discovered, there is something rather magical about a clipboard. It is, like the peaked cap, associated with authority. It commands respect and confers on its owner a degree of integrity. This can be useful as we'll see later, particularly if you encounter a shark salesman.

You will need to take a look at the underneath of the vehicle, and this may involve lying under it so *old clothes* and *something to lie on* are called for.

The tool kit is a pretty 'low tech' affair: it's surprising how little is needed. An *old screwdriver*, a *magnet*, a *torch*, a *wire brush* and a *few old rags* will suffice. A *tyre pressure gauge* is also quite useful. If you have a pair of *portable car ramps* (or you can borrow some) these are very worthwhile for looking at the underside of the car. If not, a *good jack* is handy, although there should be one supplied with the car being sold.

That's all you need to take with you, now all you have to do is to prepare yourself.

Conclusion

It is very daunting to a complete beginner to do anything for the first time whilst being watched by someone else. Car inspections are no exception, in fact, examining a car under the eye of an experienced salesman can be a little unnerving. Are you ready to carry out your examination under his critical gaze? Will he realise that you don't know how a car works? It doesn't matter much if he does realise: by the time you've checked the vehicle over, you'll probably know more about it than he does. But you will feel better about doing it if you are well prepared. So, read through the second part of the book explaining the tests at least twice, so that you get a fair idea of what you are going to do. Then, if you are doing this for the first time, it is a very good idea to carry out a complete test on a

friend's car, or even on your own. Once you've been over the whole thing once, you'll feel a lot better about doing it again.

So this is it. Now it really is time to go out into the big wide world and get yourself a set of wheels, but before you do, make certain that you are fully prepared. In particular you should:

1. decide on the sort of car you are looking for;
2. assess what the overall running costs are likely to be;
3. be realistic about what you can afford;
4. find someone to accompany you;
5. assemble your clipboard, torch etc; and
6. have a good look over a friend's car to familiarise yourself with the tests you will be carrying out.

PART II

Introduction

The second part of this book deals with the actual business of car inspection; the nitty gritty of getting down to finding any faults a car may have. It may look rather long and daunting, but don't worry. Most of the tests are simple and many of them take less time to do than they do to describe.

Part II is intended to be used in conjunction with the test sheets in Appendix 1 at the back of this manual, and as you may have noticed, each test is numbered. Each number corresponds to a number boxed in bold in the text in this section e.g. **11**. You can therefore take this book with you to refer to if you can't remember how to perform a particular test on the checklist.

The first section deals with a series of preliminaries, most of which are concerned with trying to ensure that you don't buy a stolen car, or one that is being sold by a shark dealer. Admittedly the chances of being ripped off in this way are not high, but they do exist, and it is always better to err on the side of caution. If you know the person selling the car, or you are happy that the dealer selling the car is OK, then of course you may choose to omit this section.

This is followed by a series of preliminary tests which are designed to determine whether or not the car has been looked after with the love and care it deserves. This will not necessarily tell you a lot about the car itself, but a neglected vehicle is, as you already know, not likely to be up to much. You may in such an instance decide to save yourself the trouble of a full examination.

The remaining sections show you how to carry out a systematic and detailed examination which will enable you to spot almost any fault in almost any car.

Obviously a general work of this type is intended to be used to check out an average, mid-range type of vehicle, and it is not possible to cover every single detail found in every type of car. If you are looking for something really fancy with a computer-controlled cigar lighter, turbocharged windscreen wipers, and a dashboard attempting to emulate the console of an inter-galactic space cruiser, then maybe you are after something rather beyond the scope of this book, but for most purposes it will be more than adequate.

You may feel that it is not necessary to carry out every single test suggested, particularly if you are buying a car with a warranty. This is up to you. What is offered here is a set of guidelines, which if followed, will greatly reduce your chances of buying a duff car.

First things first

For many of us the search for our car will begin with a perusal of the local 'Car Trader' type magazine. This is a good place to start as you will find a wide range of cars advertised and you won't have to travel too far when you want to look at one.

You will notice that many of the adverts are followed by the word 'Trade', or followed by the letter 'T'. It is compulsory for anyone who buys and sells cars for profit to declare that they are traders when placing an advert. Unfortunately there are in every town a number of shady characters who, whilst pretending to be selling privately, are in fact used car dealers. They usually operate from home, buying one or two cars each week from motor auctions, and, after a quick clean and polish, offering them for sale for considerably more than they paid. As their customers assume that these people are private sellers, they also assume that they do not have the same rights as when buying from a bona-fide dealer. Although the public is in law given the same degree of protection whether or not a dealer is honest, the last thing anyone wants is to become involved in a legal dispute over a second-hand car. It is far better to avoid dealing with such people, and fortunately it is not difficult to expose these rogues, as we'll see.

The first thing you are going to do when you have seen adverts for a few cars which may suit your needs is to phone up for more information. Have a notepad ready to jot a few things down so that you can make comparisons later on. Begin your call by saying that you are calling about the car. Don't specify the type of car, just say 'the car'. If the seller replies by saying 'which one?' then you know you are talking to a dealer. Refer back to the advert and if it is not a 'Trade' advert then don't pursue the matter any further.

Go on to ask if the seller is the owner of the vehicle. Anyone, other than a bona-fide dealer is required to inform the licensing authorities when they acquire a vehicle. The registration document has to be sent away so that the name and address of the new owner can be printed on it. As you are probably aware, the computerised licensing centre in Swansea does not perform quite as efficiently as intended and a new registration can take weeks or even months. This is simply too long to wait for someone whose only interest is to turn over a quick profit, so they get round the problem by claiming to be selling on behalf of someone else.

If you telephone in response to an advertisement and ask the seller if it is his car which he is selling, you might hear a heart rending tale about how the car belonged to his brother-in-law who recently died of a strange illness while travelling abroad. The other favourite is that it belongs to his mother-in-law who looked after the car with the utmost love and affection, but she is now unable to drive following a stroke.

Don't believe a word of it! The car is probably a heap he picked up in an auction last week. After a quick spray over the rust with Dupli-Colour and a good polish he'll be selling it for far more than he paid for it!

Assuming that the seller is the owner of the car, or is a genuine dealer, then the following questions are worth asking.

☐ How many owners has the car had?

☐ How many miles has the car done, and does the seller know whether the mileage is genuine?

☐ When does the MOT run out?

☐ Is there anything wrong with the car?

☐ Why is it for sale?

☐ Will you be able to test drive it yourself?

☐ Would they object to you having the vehicle examined by a professional examiner?

You cannot assume that you will receive truthful answers to all your questions, but most folk are reasonably honest and if you are told that the car has a high mileage, only a few month's MOT, or needs some repair work, then you are better off spending your time looking elsewhere. The question about having a professional examination is a bit of a trick. Although you don't intend to have the car professionally examined, if there is an objection, it suggests that the seller is aware of some fault which he hopes will not be discovered!

One thing to be very wary of is the seller who offers to bring the car round to your house. If you do receive such an offer, politely decline, and tell him you wish to see the car at the seller's premises. He may be trying to be helpful of course, but in addition to all the awful cars about, there are some rogues too, and we must always be careful. If the car being offered for sale was stolen, or an unroadworthy wreck fit only for the scrapheap, it would be very convenient for the seller if you didn't know where he lived, and were unable to find him at a later date. If the car is a heap, then you will not of course be buying it but nevertheless, it is a sound policy always to inspect the vehicle at the seller's premises.

As a general rule it is not worth going to look at a car with less than six months' MOT. A car with a full year's MOT is easier to sell, and will fetch a higher price, so anyone selling a car is likely to get it MOT'd, unless of course they have reason to believe it will fail. Find a car which has recently passed an MOT test.

Sooner or later you will locate a car which seems to be worth considering, so now at last you can begin your examination.

Preliminaries

Beware love at first sight! It can be almost as

hazardous with cars as with people, and heartbreak can all too easily follow.

Step back a little, be detached, and above all be vigilant. The best thing to do when you first see the vehicle is to form a *general impression* 1. Does it look as though it's been treated with care and attention, or does it look more like it's been used to carry cement round a building site? If it is being sold by a car dealer, it will probably be gleaming and looking as though it has never been out in the rain in its life; dealers are very good at making a car look like that. If you are buying privately, it will probably look a little less magnificent, but if it looks like it's been abused and neglected, then you can be pretty sure it has not been treated well, and is not likely to be a good buy. Neglect is far worse than old age or high mileage, so if it looks tatty, then carry out a few of the preliminary tests at the end of this section, and if these confirm your first impression, then don't waste any more time on it.

Assuming that the car doesn't look too bad, or even looks magnificent, then we can begin the examination procedure. For this it is necessary to turn your attention away from the car for a few minutes, and deal with a few other matters.

Selling cars brings out the worst in all of us, and we all know that we would be foolish to allow ourselves to be carried away by the enthusiasm, of a car salesman who is on commission. So too with a private sale. Even the most upstanding and honest of citizens can become somewhat economical with the truth when we start asking questions about a car he is trying to get rid of. Nevertheless, there are things we want to know about the car, and we can discover quite a lot, just so long as we don't place too much faith in the answers we receive.

Ask *why the car is for sale* 2, and *if there is*

anything wrong with it [3]. Ask if the mileage is genuine and whether it has been involved in an accident at any time. Ask to see the registration document and the MOT certificate.

Now is a good time to produce the clipboard and test sheets. Be nonchalant about this, but do note the reaction of the seller. As we said earlier, the use of a clipboard can bestow an air of authority onto its user. For all the poor seller knows you could be a representative of the Department of Transport, the Trading Standards Office, the Old Bill, or the KGB! If the seller gets a bit agitated, or shows signs of unease, then there is no need to reassure him if you don't want to. Be polite but evasive and simply say that you are carrying out a full inspection.

The preliminary questions on the sheet will provide a lot of useful information about the car. They will also put the wind up anyone who is deliberately attempting to sell you a duff one. He may suddenly recall that the car was involved in a very slight accident with a 38 ton lorry a couple of years back, but it was only slightly scratched! If he starts to 'remember' faults he forgot to mention when you first asked, then you are probably dealing with a rogue, and may decide to save yourself the trouble of going any further.

Note down the registration number of the vehicle, along with the make, model and year it was first registered, Write down the name of the dealer, or in the case of a private sale, *the name and address of the seller* [4].

Assuming that the seller has so far survived your cross examination, turn your attention to the registration document. If the car is being sold by a car dealer note down the name and address of the previous owner, this will be very useful later on. Note the *number of previous owners* [5], and the *date of the last transfer of ownership* **6**. If a private seller bought the car recently and is selling it again, then there is a reason for this quick change of mind. Maybe he's discovered that he was sold a heap, and he wants to pass it on to you!

As regards the number of previous owners (or keepers as they are described on the registration document), this is when you realise that a second-hand car is more often a fourth or fifth hand car. It is generally reasonable to presume that the lower the number of previous owners the better. If someone intends to keep a car for a long time, then more care and attention is likely to be lavished on it and there is nothing like regular maintenance to ensure a long and healthy life.

Note down the *vehicle body* **7** and *engine* **8** *numbers* as they are recorded on the registration document. Ensure that there is no record of the vehicle being *an insurance 'write off'* **9**. It used to be the policy of the licensing authorities to make a note on the registration document if a vehicle has been an insurance write off. If the car was rebuilt, the record remained and a buyer could check what he was buying. This practice has been discontinued, but there is some debate going on and it is possible that the practice of recording write-offs will be re-introduced. It is therefore worth checking to see if there is such a note on the log book.

Now turn your attention to the MOT certificate. Many people assume that because the car has an MOT then it's roadworthy. Unfortunately this is not the case. All a test certificate means is that the vehicle was roadworthy on the day the test was carried out and it is legal to drive it on a public road, provided that it doesn't become unroadworthy in the meantime. A lot may have happened since the test and unless the certificate is only a few days old, it doesn't guarantee much. A recent random MOT test on a batch of new cars produced a fair number of failures, so it is always worth carrying out a full examination yourself.

Note the *mileage recorded on the MOT certificate* **10** at the time of the last test, and compare this with the *mileage recorded on the vehicle's mileometer* **11**. The car will have a higher mileage than the one shown on the certificate, even if it is only a few miles clocked up on the return journey from the testing centre. Unless of course the mileometer has been tampered with. If the mileage recorded on the vehicle is mysteriously less than the figure on the MOT certificate, then you need look no further at this one. A professional shark will not be caught out like this as he will simply arrange for the car to be re-tested after he has wound the clock back.

However, there are amateur crooks around who may overlook that small detail.

If all seems well so far then we actually begin the examination of the car itself. The first thing to do is to have a look under the bonnet [12]. The bonnet release catch is located in the car, usually somewhere under the dashboard. There seems to be some sort of competition going on between car manufacturers to see which one can produce a car with the most difficult bonnet release catch to locate. So unless you enjoy having to grope around under the dashboard of a car it is best to ask the seller to lift the bonnet for you.

If this is your first close encounter with the internal clockwork of a motor car, then you may find it a little bewildering. It may look like a confusing mass of metal and other bits and pieces, but it all has meaning and purpose. Everything has a part to play to make the thing go. Don't worry! All will become clear in the fullness of time, and you'll soon get used to it all. The engine is the big hunk of metal in the middle which is surrounded by various ancilliary components. If you are not used to poking around amongst the innards of a motor vehicle, then do be careful. If the engine has been running recently, then parts of it will be more than hot enough to scorch human flesh!

Once again, the first thing to do is form a *general impression* [13]. This is difficult without some experience of what things should look like. Nevertheless neglect shows here as much as anywhere else. The engine compartment will not often look clean and shiny, but it should not be so filthy that it looks like it recently had a prolonged encounter with *an oil slick* [14]. A dirty engine does not mean that it is no good, but it does indicate some degree of neglect which is never a good thing. You'll not often come across an engine that doesn't leak any oil at all, so don't expect perfection. On the other hand you don't want to end up with a car that is leaking a lot: oil is far too expensive for that. Oil leaks are often an indication of poor workmanship during repairs or maintenance, so a leaky engine is something it is best to avoid.

Now look for the *engine* 15 and *body numbers* 16. These are usually stamped onto aluminium plates which are attached in a prominent place. Normally the engine number will be near the top of the engine, whilst the body number will be at the rear of the engine compartment. You will probably need to wipe them over with a rag to make them legible. Copy the numbers down onto the test sheet, and make sure that the numbers are the same as shown on the registration document. If the body number is different, then you are not looking at the right car! If the engine number is not correct, the engine has been replaced at some time. There is nothing wrong with that in a car which is getting on a bit; major transplant surgery is often used to prolong the life of a vehicle. However, the change should have been reported to the licensing centre, and the registration document altered accordingly. It is an offence not to do so. If the engine had come from a stolen vehicle and this were ever discovered, you could find yourself the owner of a car with no engine!

Having determined that the car you are looking at is the car it is claiming to be and that the engine does belong to it, you now need to find out how well it has been looked after recently. Ask the seller if he has any *records of servicing and repairs* 17. A car which has been serviced regularly from new will be in a much better state than one which hasn't.

By far the most important element of vehicle maintenance is to ensure that the engine contains sufficient oil at all times and to change it regularly as specified by the manufacturer. This will prevent the engine from wearing out prematurely; any engine which is used with insufficient lubricant will wear out very quickly. So, the first thing to do is to check the oil level. In order to obtain an accurate reading, the car must be fairly level, so if it is parked on a slope it will have to be moved.

About half way down one side of the engine you will find the dipstick. This is a long thin piece of metal, often curved into a loop at the top to make it easy to withdraw. Remove it, wipe off the oil with a rag or tissue, then replace it. Take it out again, and note *where the oil comes up to* **18**. It should be between the max and min marks on the dipstick. If it is at or below the minimum level, you can assume that the car has not been well maintained. Topping up and changing the oil is one of the simplest tasks and if the seller has not managed to do that, then he certainly won't have managed to do anything else. A badly maintained car is likely to be a liability, so you would be as well to steer clear.

It should be pointed out that the condition of the oil is as important as the oil level. Modern lubricants will withstand the rigours of life inside an internal combustion engine for a considerable time but even oil wears out, and must be replaced at regular intervals. The *condition of the oil* **19** is a little difficult to ascertain without some experience; it always looks pretty dirty, even when it has been recently changed. However, the oil on the dipstick should (despite being black) have a slight translucency about it when it is warm. If the engine is cold at this stage, then it may be worth re-checking it after the road test.

Now *look at where the oil is put into the car* **20**; the filler cap on top of the engine. This is often identified by having an advertisement for a particular brand of oil in it. Remove the cap, and examine its underside, and the sides of the hole it occupies. Make sure that it is not coated with a creamy white deposit. If it is, then water is somehow getting into the oil, a problem which will indicate something fairly serious like a blown cylinder head gasket, or something very serious like a cracked engine.

Replace the oil filler cap, and look now at the radiator. If the vehicle doesn't appear to have one, it may be because it has an air-cooled engine, but these are rare these days, so you will probably find the radiator at the very front of the engine compartment. It is worth checking the level of coolant in the radiator, not because the water level tells us much about the car, but because it indicates whether the last owner looked after the car properly. Firstly, place a hand near the top of the radiator. If it is hot then do not under any circumstances remove the filler cap. The cooling system works under pressure, and removing the top will have the same effect as removing the lid

of a pressure cooker before it has cooled down: a jet of boiling water will shoot out and fall on anyone within range. There will be plenty of time for the engine to cool while you look at something else, so you will have to do this one later. If the radiator is cool, or hand hot, you can safely remove the top and *check the water level* [21]. The water should fill it to the top, or in the case of a cooling system with an overflow, it should come up to the mark on the plastic container near the radiator. Look carefully to see if there is *any oil floating on top of the water* [22] – oil getting into the water can be as bad as water getting into the oil. If it is winter, there should be *antifreeze* [23] in the water which will be obvious due to the coloured dye it contains.
Somewhere, sticking out from the side of the engine will be *the oil filter* [24]. This is a bright coloured cylindrical object which looks a little out of place among the drab metallic colours of the rest of the bits and pieces under the bonnet. This should be replaced regularly when the oil is changed, so if it looks like it's been there for years, e.g. the casing may be going rusty, then a lack of maintenance is indicated.

The battery [25] should be examined next: that's the thing that always goes flat on winter mornings. It will not look flat however: a battery always looks much the same whether it is full of the joys of life, or is stone dead. What we can tell from looking at the battery is once again whether the previous owner has looked after the car properly or not. Carefully remove the top so you can look inside it. Care is needed here, as the liquid it contains is sulphuric acid at a strength great enough to damage clothing, or cause skin burns. Inside the battery you will see some rows of metal plates. These should be covered by the acid and not showing through. The battery is simple to maintain, and a careful owner will give the battery an occasional top up.

Make sure that *the tyres* 26 are alright and have at least 1 mm of tread left and no bald patches. Again we are checking up on the owner here, as no sensible person drives around with worn tyres. If you want to go the whole hog, check the tyre pressures. The two front tyres and the two rear ones should be at the same pressure. If one is a little low then it hasn't been checked recently. If any of the tyres are badly worn on one side but not the other, then there may be a problem with the steering which will need to be adjusted. However, caution is required, *as uneven tyre wear* 27 can be an indication that the car itself is twisted, either due to it having been involved in a serious accident, or possibly it's one of those notorious concoctions made up from more than one car.

By now you should have a good idea of how well the car has been looked after, and you'll probably have a good idea of whether the person selling it is reasonably trustworthy. If you are in any doubt about what the vehicle is like, then you may be better spending your time going to look at something else – there are plenty of other used cars around. If so far things look fine, then the car has already had a more thorough inspection than a lot of used cars get before they are bought. But this is only the beginning, a preliminary once over. The following sections describe in detail how to carry out the full inspection.

The exterior

The most important aspect of a used car is not, as many people think, the condition of the mechanical components, but the condition of the bodywork. It is always possible (albeit rather expensive) to replace a worn-out engine or gearbox. The recent upsurge of interest in transplant surgery by the medical profession is after all only an offshoot of the car repair business which has been carrying out transplants since motorcars first arrived on the scene! However, once the bodywork of a car has deteriorated past a certain point, then there is little that can be done to restore it, and the once shiny car becomes a heap of metal which can be used only for recycling, and as a donor for an occasional spare part transplant.

So we will in the next two sections be paying a lot of attention to the condition of the bodywork of the vehicle, firstly from on top and later from underneath. The main object of our investigation will be to look for evidence of rusting and general deterioration. We'll also be looking for signs of accident damage and repair work.

Later on in this section, the condition of the steering and suspension system will also be looked at; not because it is part of the bodywork as such, but it is more convenient to do this during the course of the examination of the exterior of the vehicle.

The first thing to determine is whether the car has been *re-sprayed recently* **28**. Not many cars are repainted during the course of their lives due to the very high cost of this type of work, but you do occasionally come across one. If it has been done, the seller will almost certainly point this out to you in the hope that it will encourage you to buy it. Buying a re-sprayed car is not something I would advise anyone to do, unless you saw it before the work was carried out. This is because you cannot see what is underneath the shiny new paint and although it all looks resplendent for a while, there may be all sorts of poor repair jobs underneath which will show up as rust patches before too long. A re-spray is a very expensive undertaking and it is unlikely that anyone will pay out that much money unless what was underneath was in a pretty poor state.

The experienced eye can tell at a glance whether a car has been re-painted since it first left the factory. This is because the paint used is quite different, the original being a very high quality coating which has to be baked on in an oven, whilst-respraying is carried out using an inferior air-drying paint. However, you don't have to be an expert to tell the difference as there are several tell-tale signs to look for.

Firstly, no car is going to be perfect and there will always be a few minor scratches and marks on the paintwork. If the vehicle is a few years old, yet shows no signs of wear and tear, then you can be pretty sure it's been re-painted. Immaculate paint work on a used car is simply too good to be true and there is likely to be some rust lurking underneath which will manifest itself before too long. The other tell-tale sign of a re-spray is 'overspray'. When paint is applied using a spraygun, it is blown out in a fine mist which gets everywhere. Although it is possible to cover up, or 'mask' the places where a coating of paint is undesirable such as the windscreen and bumpers, it is very difficult to ensure that the paint goes only onto the places it is required. There will always be some overspray somewhere, on some of the chrome fittings, or the rubbers round the doors, the back of the numberplate and so on. If you suspect a re-spray, your suspicions will probably be confirmed by signs of paint where you wouldn't expect to see any. However, do not assume a complete re-spray has been carried out solely on the evidence of an isolated bit of overspray. This is more likely to indicate a body repair which has been carried out recently, which is something that will be dealt with later in this Part. Before we get down to the detailed look at the bodywork, it is worth taking a few steps back to see what the car looks like from a short distance away.

Look at the vehicle in relation to the surface on which it is resting. The *bumpers should be parallel to the ground* 29 and the sides of the car should also look straight in relation to the road. If it looks like you'd expect it to look with a 24 stone driver sitting in it, despite the fact that it is empty, then the *car will somehow be twisted* 30. This may be associated with uneven tyre wear as suggested earlier. Such faults may be due to there being something wrong with the suspension, but it is unusual for uneven wear to occur. It is more likely that the car has been well pranged at some stage and is twisted as a result. Although there are people who claim to be able to straighten out twisted cars, it's pretty dodgy, and its a much better idea to go and look at another car.

Security numbering 31 of vehicles is now very popular, so if the vehicle has been done, check that the number which is etched on the windows is the registration number of the car and ensure that the same number appears on all the windows. If the back numbers are different from those on the front windows, or there are no numbers on either the back or the front, then you may be looking at two cars joined together. Having said that, if anyone is going to go to the trouble of re-building crashed or stolen vehicles, then it is not a great deal more trouble to obtain some second-hand, unmarked windows from a scrap yard and fit them to the vehicle. A serious re-builder would probably etch the correct numbers onto the glass, so don't assume anything just because a car has the correct numbers etched onto the windows. The main benefits of security numbering are reaped by the people who make a living from doing it!

Still on the lookout for signs of a car that has been bent at some stage, *open each door in turn* **32** and shut it again. Ensure that it opens and closes easily, and that it fits properly into its recess. Try *the boot and bonnet* **33** in the same way. Now crouch down at each corner of the car and look along the length of *the bodywork* **34**. Any repairs which didn't look so obvious before will show as a distortion of the smooth lines of the bodywork. Check across the back and front in the same way. Hopefully the vehicle will not have been resuscitated following a major accident, so we can now look for signs of repairs necessitated by minor bumps and rusting of the bodywork.

Most cars are laid to rest as a result of being eaten away by rust. Despite the voraciousness of this affliction it is not always easy to see as it eats away at the hidden corners and recesses of the vehicle, appearing on the surface only after much damage has been done.

Sometimes rust holes are repaired using the body repair kits available from motorists shops. These are fine, provided that a proper repair is carried out, but almost invariably an amateur job (and many professional jobs) will sooner or later begin to rust again. Amateur body repairs are usually easy to spot. It is a skilled job and repairs done by someone who isn't an expert will somehow lack the finesse of a professional restoration.

Since water travels downwards, cars will usually rust from the bottom up, so we will begin the search low down. The long metal strips which run the length of the car below the doors are *the sills* **35**. On many cars they form part of the structure of the vehicle, so that the strength of the vehicle depends to some extent on the sills being in good condition. If the sills are weakened by rust holes, MOT failure is likely. A repair with a fibreglass body repair kit is not satisfactory for MOT purposes and the only remedy is to have the sills replaced, a job which is not easy to do and usually costs a lot to have done. This does not seem to deter a seemingly large number of car owners from repairing sills with body repair kits, either through ignorance, or in the hope that they will be able to fool the MOT examiner or the person they hope to sell the car to. The MOT tester will certainly spot such repairs and so will you. All that is needed is vigilance.

So, have a good look at the sills. They may be covered with a layer of mud, in which case it is worthwhile scraping it off to have a look at what is underneath. Sills often rot away from inside, so what may look like a bit of rust on the surface may in fact be the final stages of the metal being eaten away. If you do see rust spots, then examine them carefully. Ideally it is best to prod at anything which looks a bit doubtful with a small screwdriver to see if it goes through. This will not necessarily go down too well with the seller (if you start poking holes in his car, he might get a little upset) so you'll have to do the best you can. With a little bit of practice you'll soon learn to tell the difference between a bit of surface rust and a car which is being eaten away from within.

Any *repairs* to the sill **36** which have been made are usually fairly easy to see; the surface of the metal will have awkward looking bumps in it and the colour of the paint won't quite match. Unfortunately, this is not always so obvious as far as the sills are concerned, as they often look a bit grotty anyway and it can be difficult to spot differences in paint colour when there is a layer of dirt masking the differences. However, this is where the magnet becomes very useful. Although a magnet will be attracted to the steel from which the original bodywork was made, it will show no interest at all in the plastic substance from which body repair kits are made. It is a simple job to go along the vehicle with the magnet to see if there are any areas to which the magnet seems to be unattracted. The sills are more likely to rust at the lowest points and at the ends, so check carefully here.

If the car you are looking at is one of those rare specimens with an aluminium or fibreglass body, then the magnet will not be attracted to it, but if this is the case, you will not have to worry about rust at this stage. However, even rustproof bodies are attached to a steel chassis which is prone to rust and will have to be looked at later on when we deal with the underneath of the vehicle.

Now look at the doors, especially near the bottom.

These too have a habit of rusting from inside due to water seeping in past the windows, so if the *doors are rusty* 37 at the bottom, it probably goes right through. On older vehicles this is a common problem. It is unlikely to cause MOT failure unless there are sharp metal edges around the hole which could be dangerous. It is alright for these holes to be patched with repair kits as far as the MOT is concerned, but, as the rust is usually eating away from the inside, repair work on the outside of the door is rather temporary, and it is not long before the rust re-emerges through the paintwork. Use the magnet here to check for repairs which may have been done.

Go over all *the bodywork* 38 checking for rust or repair work. Give particular attention to the nooks and crannies which are particularly prone to rust. The wings, or areas of bodywork near the wheels and around the headlamps are particularly vulnerable. Use the magnet to check for repairs. The corners of a car tend to suffer most from minor accident damage, so you may find evidence of repair work here. Minor accident repairs are not as much to worry about as patching up of rust holes. This is partly due to the fact that a repair will normally last longer in a place where rust does not have a hold, i.e. a dent caused by a minor collision. It is also because accident damage is usually repaired by experts since the work is often paid for by insurance companies.

It is also necessary to look for rust *under the bonnet* 39 to see what the bodywork looks like here. It is from this vantage point that we can see the inside of the wheel arches. These are prone to rusting due to mud sticking to the underside which traps moisture. Any holes or signs of serious corrosion should be taken seriously, as on many cars some part of the steering or suspension system will be attached here. Any weakness due to rust will not only cause MOT failure, but could prove highly embarrassing and dangerous if it were to fail at an awkward moment causing the suspension to collapse and control of the vehicle to be lost.

Provided that the bodywork of the vehicle looks reasonable, we can now take a look at some other aspects of the car, which for want of a better description come under the loose heading of 'exterior'.

The car should have a jack supplied with it, along with a wheelbrace, which is a spanner to undo the wheel nuts in case of a puncture. Somewhere the car will have some jacking points. These are the places under which a jack may be placed in order to raise the vehicle. They are usually located under the sills, behind the front wheel, and forward of the rear wheels. Ensure that the vehicle is on a reasonably level surface and jack up one corner, so the wheel is clear of the ground. It is advisable to place a brick in front of and behind one of the wheels which is still on the ground so the car doesn't attempt to go anywhere while you are busy pursuing your investigation. It is a strange irony that *the jacking points* **40** which are specifically there to take the weight of the car while it is raised clear of the ground are usually the first things to rust away, so it is as well to have a good look at them before you carry out this procedure. If they have gone it is still possible to raise the vehicle by placing a jack under a strong part of the underside – look for a thick piece of the supporting structure near one of the wheels.

With one of the wheels raised clear of the ground look at *the condition of the wheel arches* **41**. This is the cavity which surrounds the wheel and is normally to be found with a generous coating of mud, grime and all manner of other unpleasant substances which are perhaps better not mentioned. In order to inspect the state of the metal, it will be necessary to remove some of this muck. It will fall away quite easily if prodded with the screwdriver, so get rid of as much as you can, particularly in all the awkward corners where rust will so easily take hold. On older vehicles you will often find some rust here, but you want to ensure that you don't end up with a car which is rotting away. So, if you see any signs of rust, poke it with the screwdriver. If there are any holes, or the screwdriver goes through, then the rot has set in and will require remedial work to prevent it from spreading.

On some cars you will see a large spring behind the wheel which somehow connects itself to the top of the wheel arch. This as you will probably realise is part of the suspension system – that curious arrangement of springs, dampers and 'boingy things' which somehow manages to make a journey along a bumpy road feel reasonably smooth. It is important that the place where these connect onto the body of the car is strong enough to take the load and is not weakened by deterioration of the bodywork, so examine these areas with particular care.

Now turn your attention to the wheel itself. You've already checked the tyres for wear, but there are plenty of other things to have a look at. Place one hand on each side at the 3 and 9 o'clock positions and waggle the wheel to see if there is any play, or looseness. The front wheels will of course move from side to side, steering the car would be awkward if this were not the case, but the steering movement is slow and difficult. What you are looking for here (but hoping not to find) is a wheel that can be waggled a little. Repeat this waggle with the hands in the 6 and 12 o'clock positions, i.e. with the hands at the top and bottom of the wheel.

In the case of a *rear wheel* **42**, then any play will normally indicate a worn wheel bearing. This may simply need to be adjusted, but may need to be replaced.

Front wheels **43** are rather more complicated. Any play could indicate wear in any one of a number of different places, some of which are fairly simple to put right. However, some steering faults are very expensive to repair and due to the wide number of variations between different cars and the relative degree of complexity of the steering system, it is not possible to devise a system of tests which will enable anyone without expert knowledge to find out what is wrong. This is no great problem however.

There are a few aspects of motor vehicles on which expert advice is available free of charge, and steering problems fall into this category. All will be revealed later in the book, so don't worry at this stage: it can be looked at again later if the car is otherwise alright. Just to make absolutely sure the steering gear is not worn, repeat the 'waggling' tests with the steering wheel turned fully to one side – wear will show up more on full lock.

While you have the corner of the car raised, you can carry out all the checks detailed in the next section which deals with the underneath of the vehicle. Do each quarter of the underside in turn whilst the car is raised, but do be careful. It is unwise to rely on the jack to hold the car up. If it were to fail while you were underneath you would experience first hand how heavy the car is – quite heavy enough to break bones if they happened to be in the way! Before you go under, a pile of wooden blocks or good quality bricks should be placed in such a way that if the jack were to fail, the weight of the car would fall onto them, and not onto you. Better still, if they are available, is a pair of axle stands.

After you have lowered the car again, there is one more thing to look for, and that is *wear in the suspension damping system* **44**. Although the suspension system uses springs of various types to absorb some of the bumps as you motor along, the use of springs alone would produce a very bouncy ride, which, quite apart from anything else, would cause a very high incidence of car sickness. To counter this effect a damping system is used, which allows the spring to absorb the bump, but does not allow it to continue to 'boing' up and down. These dampers do wear out, so it is worthwhile testing them. To do this, place all your weight on one corner of the car above the wheel. Release the weight suddenly and watch what happens. The car should quickly rise to its former position, with only a very small hint of a bounce afterwards. If it seems to continue to bob up and down after you release your weight, then the damper is worn. It is unusual to find a vehicle with more than one corner affected, so you will be able to make comparisons. If one corner bounces more than the others, then there is a duff damper.

There are a few cars around with somewhat more sophisticated suspension systems which only operate when the engine is running, so if the car feels very solid with no give in it when you put your weight on it, then you are probably looking at one of those. This test will therefore not work in this instance. If you are buying a car at what is politely known as the 'lower end' of the market, then this type of car may be better avoided altogether. The more sophisticated the car, the more it costs to fix when it goes wrong, and it is a fact of life that older cars go wrong more often.

That's it as far as the exterior is concerned. The next step is to have a look at the underneath.

Underneath

Old cars never die, they just get towed away to be recycled. Most cars making their final journey do so because they have succumbed to the ubiquitous rusty chassis syndrome: the killer rust strikes again!

There are basically two ways of building a car; the first is to construct a steel chassis, which is a strong framework on which the rest of the car sits.

A more modern method is to use the monocoque construction where the body is designed so that it is strong enough to take its own weight and that of the engine etc. This type of structure is widely used by the beer industry – as you know, an empty beer can is easily crushed in one hand, but if it is placed upright, it will take the weight of someone standing on it. This second method clearly uses less materials and produces a lighter car, but there are advantages to a chassis construction method, so many cars are designed as a sort of compromise, with some sort of steel framework to strengthen a monocoque type body.

Engineers have devoted their lives to finding the best designs, using all the modern aids available in this computer age. Rust however is indifferent to all this. It doesn't matter how fancy or clever the design, if it's made of steel, then it is prone to rusting.

Once rust gets a hold it spreads relentlessly corroding the metal and eating away at the supporting structure. Every year, the MOT test weeds out such cars, declaring them unfit for the road and condemning them to the knacker's yard.

People complain about this, but it has to be done. A car which is seriously weakened by rust is a danger to its occupants and other road users. At some point, something will give, and control of the car will be lost. This usually happens at a time of greatest stress, e.g. whilst cornering, or when travelling at high speed.

It is sometimes possible to carry out repairs to the underside by welding patches on, or replacing a section of the chassis. Occasionally this can be done without too much difficulty, particularly on cars with a known weak spot. In such cases where it is known that a car is prone to failure at a particular point, a new part is available to weld on.

Having said that, once the rot sets in, it is the beginning of the end and it is far better to find a car which is in good condition underneath than one which will need to be worked on in the forseeable future. In order to ensure that all is well on the underside you will have to take a look at it and in most cases this means that you are going to have to get under the car to do this.

Although there are people who seem to derive so much pleasure from crawling around under motor-cars, that they spend all their spare time doing it, there is a prevailing opinion amongst most ordinary folk that the underneath of a car is a place to avoid at all costs. As you will soon discover, this is not the case at all. The underside of a car is extremely interesting; another world which exists outside the experience of most of us. Added to this is the sense of achievement gained from conquering the fear of the unknown and the satisfaction gained from doing something for oneself.

If the car is being sold by a dealer with workshop facilities, then you will have to deny yourself this opportunity of broadening your experience. Tell him you wish to inspect the underside of the car and ask him to raise it up on his car lift for you to do so in comfort. He might grumble a bit, but if he wants to sell the car, he's likely to oblige.

If you are involved with a private sale, then you are not likely to be able to look at the underside in quite this degree of comfort. Some cars have a fairly high ground clearance, whereas others, particularly those with small wheels, have very little room underneath. If this is the case, or if you are of rather generous build, then it will be necessary to raise the car to allow more room. There are several ways of achieving this.

As previously mentioned, the underside can be

looked at while you have got a corner of the car jacked up. You don't have to actually get right under the car (though there is nothing to stop you from doing so if you want to). All that is required is to get enough of yourself under so you can see the underside and poke around a little. It will be sufficient to get your head and one arm under. If you do this, do be very careful not to leave bits of your body sticking out where they may be run over by passing vehicles!

The best way of getting under a car is to use a pair of portable ramps as they give you plenty of room. If you feel that your interest in cars has been so stimulated by carrying out your own inspection that you may in the future extend this to carrying out your own repairs, then they are a worthwhile investment. If you do use ramps, be sure to 'chock' the wheels which are still on the ground to prevent the car from rolling away.

If neither jack or ramps are available, you can raise one side simply by driving two wheels onto the pavement. Strictly speaking it is illegal to drive on the pavement and I do not of course wish to encourage anyone to do so, so I'll leave this one to your conscience.

Once you've arranged things so that you can have a look at the underside, then it's time to take the plunge. Before you do dive under the car, have a careful look at the surface you are about to lie on. There are all manner of unpleasant and disagreeable things deposited on public highways, some of which it is better not to lie down on! It is worthwhile wearing an old coat or jacket just in case.

The main purpose of the underside examination is the search for rust. It used to be said in the 60's that car manufacturers sprayed rust onto cars prior to painting them. Things have improved a lot since those days and the ability to withstand corrosion is now one of the main selling points of new cars. Although you won't find a used car without a bit of rust somewhere, the object of the excursion into the vehicle's nether regions is to find out if there is any serious rusting going on. It isn't always as obvious as one might expect – it may be hiding under a layer of mud in some forgotten crevice, so you will need to be vigilant.

New cars leave the factory with a protective coating on the underneath which is called the *underseal* **45**. This is a matt black rubbery paint which will resist water and rust for many years. But small stones thrown up by wheels will chip away at this leaving small areas unprotected. This would not be so bad were it not for the problem of salt. Salt, a substance once considered so valuable that Roman soldiers were paid with it, is liberally scattered on our roads every winter during freezing weather. This substance has three main properties: firstly it makes fish and chips taste palatable, second it melts ice; and thirdly it makes cars rust very quickly. Those little blemishes in the underseal will be attacked by salt and then the rust really gets going.

Shine your torch around into any dark corners, and *look at all crevices* **46**, pipes and other unnameable things. If the car has been running recently, beware of hot exhaust pipes! Every car looks different from underneath, some are almost flat, others have a chassis and some have a mixture of the two. The chassis consists of strips of rectangular shaped metal (box sections) which may run the full length of the vehicle, or just along part of it.

The bodywork of the vehicle has somehow to be connected, via the suspension system, to the wheels. Even a car without a chassis will have some sort of strengthening in the vicinity of the wheels so that they can be joined to the rest of the vehicle. This will usually consist of *some box section steelwork* **47** attached to the main bodywork. It is obviously vital that these areas are not weakened by rust; any corrosion will not only lead to an MOT failure next time round, but could be dangerous.

Box sections do have a rather interesting little trick which is well worth knowing about. As it is made from square sectioned steel tubing, it is hollow and water can therefore get inside. Once inside, the corrosion process will begin and whilst all looks well from outside, and the coating of underseal sits there looking fine, the box section happily rots away from within. So although all looks well, the chassis on an older vehicle may be in a rather weakened state and have very little strength left. MOT examiners know all about this, and if they suspect weakness in any structural parts of the car, will have no qualms about laying into it with a hammer. If a hole appears it's a failure!

The owner meanwhile will have been elsewhere at the time, blissfully unaware of such attacks on his beloved car. It is therefore unlikely that he will take kindly to you assaulting his vehicle in the same way. Besides, it is not a good idea to do anything quite so violent unless you are sure of what you are doing; you may hit something you didn't mean to and cause damage. So if anything looks even a bit dodgy, you will have to content yourself with vigorous poking with the screwdriver.

It's a bit of a grubby job, but if you want to be certain that the car you buy is a good one, then the whole of the underside should be inspected, probing every nook and cranny in the search for rust. Look carefully at anywhere that has accumulated a deposit of mud. This will trap moisture and ensure that the rusting process continues unabated. Use the screwdriver to dislodge these deposits so you can see what's hiding underneath. It is worth ensuring that you are not directly underneath when you do this, otherwise you will get a faceful!

Be on the lookout *for signs of repairs* **48**, particularly on the box sections, Some people do actually attempt to repair box sections with body repair kits in the the futile attempt to hide the rust from the MOT examiner. He will not be fooled, and neither will you. If you are suspicious, then use your magnet to make sure that you are at least looking at metal.

Any areas which appear to have been freshly painted with underseal should be regarded with suspicion. If it is painted over a rusty area, it will only assist and speed up the rusting process. It may have been used to cover up a bodged repair, so look for signs of body filler. A properly welded repair may have been carried out and undersealed to prevent further corrosion. This is fine provided it has been done properly, but it is not a very good sign. Once a car starts to rot on the underside it is usually the beginning of the end and repairs will be nothing more than a temporary postponement of the inevitable.

You'll need to inspect the whole of the underneath of the vehicle in this way, prodding away, checking with the magnet, and giving the car a very thorough inspection. Whilst you are at it, look for evidence of a *join or seam across the width of the car* **49** which could indicate that it is made up of two cars joined together.

Although you are unlikely to find a car with no rust at all underneath, remember that what you are looking for is rust that has got a hold and is likely to weaken the structure. Experience is helpful here, but you'll soon get the hang of things and be able to tell the difference.

If the car is suffering from corrosion, then you are looking at something which will soon be on its way to being recycled. A rusty car is potentially dangerous – at worst it could cost you your life and it would at best be an expensive mistake. *If in doubt, don't buy it.*

If the underside has withstood the onslaught of your probing and poking, and all looks well so far, then we can carry on. Before you return to the land of sunshine and blue sky, there are one or two more interesting things to look at underneath.

If you look at the back of one of the wheels, you'll see a thin pipe coming out of it, which travels towards the front of the car. This is used to carry hydraulic fluid to the brakes and has to withstand a fairly high pressure. If the pipe becomes weakened by rust, then this will lead to MOT failure and sooner or later, will lead to brake failure if not attended to. Brakes have a tendency to fail at the time when they are most needed which is during an emergency stop, so it is important to make sure that they are in good order. Provided that the car is in good condition underneath, then *the brake pipes* **50** will probably be fine. Nevertheless, thoroughness is always worthwhile, so follow the brake pipes along checking for rust. They will probably be a little muddy, so clean off any areas which are suspect. Look particularly at points where there are clips which fasten the pipes to the car, as this is where rust will probably start.

You will probably come across another thin pipe while you are underneath. This is *the pipe which conveys petrol* **51** from the tank to the engine. It is less likely to suddenly fail as it isn't pressurised, but petrol leaks are expensive as well as dangerous, so have a good look at this also.

Talking of *the petrol tank* **52**, this is worth taking a look at too. It's usually at the back of the car and can be mistaken for part of the bodywork of the vehicle. If you tap it there will be a hollow sound, depending on how much petrol is in it. Not many people sell their cars with a full petrol tank, so it will probably sound very hollow. Make sure that there are no leaks in it and that it isn't badly rusted. Unfortunately, petrol tanks have a neat little trick of rusting through on top where you can't see it. There isn't a lot you can do about this, except curse when it happens. You first notice it when you fill the tank and wonder why there is a strong smell of petrol for a while until the level drops and it stops sloshing out.

The next interesting thing to turn your attention

to is *the exhaust system* [53]. There is usually one of these, but larger cars will sometimes have two. It consists of a metal tube, about two inches in diameter which goes from the engine compartment to the back of the car. Somewhere along its journey it will go into a metal box which resembles an overgrown tin can. Some systems will have two or three of these tin cans between the front and the back. Unfortunately the resemblance to the proverbial baked bean can does not stop at the physical appearance. They are made of thin low grade metal which is not very durable, and they don't last very long. The exhaust system will often look rusty, so the best you can do here is to see if it is about to fall apart. The pipes will normally outlast the tin cans, so have a look at these first. If it looks dodgy, give it a poke to see of there is any metal left.

As the exhaust system is a component with a limited life, the fact that it may be falling apart does not have any bearing on the car as a whole. If it is in need of replacement, the vehicle may still be a good buy. However, you should be able to negotiate a price reduction to cover the cost.

You may come across a car fitted with a stainless steel system. This will last the life of the car, and although initially expensive, is cheaper in the long term. If you do come across a car with one of these rather rare features, then it indicates that the owner looked after the long term interests of the car which is obviously a good sign.

There are a few other things worth a quick look while you are there. Look at the underside of the engine for *signs of oil leaks* [54]. There will always be some oil here, but if it is really dripping and oil has sprayed out to cover the underside of the car, then it is losing an excessive amount. Although a coating of oil on the underside will offer some protection from rust, oil is expensive and fixing an oil leak can be a very major undertaking and consequently rather costly. Don't buy a car which is leaking badly.

There are other places to look for leaks. Check the back of each wheel. Hydraulic fluid can escape if there is *a fault in the braking system* **55** and this will be seen at the bottom of the round plate at the back of each wheel. While you are looking at the wheel from behind have a look at the *side of the tyre for cuts* **56** and damage that cannot be seen from the normal viewpoint.

As far as the underside inspection goes that's all there is to it. It may sound pretty horrendous the first time you read through it, but there is really very little to it. Once you get the hang of it, a tour of the underneath will reveal all it's hidden secrets in less than five minutes.

It's time well invested and is probably one of the most important aspects of the examination. You can replace most parts of a car, but if its gone underneath, then it's had it.

So, come out into the world of the living. If everything looks OK so far, then you can now actually get in and have a look at the interior.

Inside the car

So at last we come to look at the interior of the vehicle. We're not quite ready to zoom off in it yet – the road test is in the next section and before we become too interested in the degree of comfort afforded by the driver's seat, there are just a few other matters which need to be considered first.

As you will probably realise by now, a general impression of the overall state of affairs will indicate how well the car has been cared for. *The condition of the interior* **57** is often a better guide than the outside as to the sort of life the car has led. Although it is possible to respray the exterior paint work, it is not easy to make the upholstery look like it has never been sat on when the car has been used a lot. If a car has a low recorded mileage and yet the interior looks tatty, then the car may well have covered a greater distance than that indicated by the mileage recorder. If you find that the state of the interior doesn't quite match what seemed to be an immaculate exterior, then maybe the car has been resprayed, and you need to take a closer look at the paintwork.

If it is possible to lift up the carpets, then look over the state of the floor of the car. If a *car floor* **58** goes rusty, it usually rusts from the inside due to water leaking past the rubber seal around the windscreen. So check that *the floor is dry* **59**, particularly in the corners. If in any doubt, have a poke around with the screwdriver. A rusty floor isn't a disaster, it can be repaired and may need to be welded to pass the MOT, so it may be acceptable on a cheaper, older car. However, corrosion does mean that water is getting in somewhere and these leaks can be very difficult to locate. Often the only way to stop a windscreen from leaking is to take it out and re-fit it. While you are looking under the carpet, be on the look out for *a welded seam* **60** across the width of the car which will be an indication of two cars being welded together to form one. You don't often find this, but it's as well to be on the lookout.

Now try out *the driver's seat* **61**. It may seem like an unimportant point, but if you are going to spend much time sitting in it, you might as well take comfort into account; ensure that it isn't too excruciating to sit in.

Turn on the ignition. Depending on the type of car, there will be some indication that the ignition is now on in the form of warning lights on the dashboard. Some car manufacturers have gone to rather extreme lengths when it comes to warning lights, providing lights for every conceivable activity likely to take place within the workings of a motorcar. Quite why anyone needs to have a light to tell them they haven't fastened their seat belt, or that they are driving around with the brakes on seems a little odd to me, but people obviously like to have lots of pretty lights, so car makers build them in. In the old days of motoring, the prevailing wisdom suggested that two warning lights were actually necessary, and they are still the only really important ones, so these are what will be considered here.

The first and less important of the two is the *ignition warning light* **62**, usually marked 'ign', and red in colour. When this lights up, it indicates that the battery is being discharged, and will go flat if it continues to discharge. Normally the generating system will ensure that the battery is charged up while the engine is running, so if this light comes on with the engine running, then there is a fault in the generating system which should be put right without delay.

The other warning light is the *oil pressure light* **63**, which may be any colour and is usually marked with a picture of an oil can, or the word 'oil'. This is designed to light up when the oil pressure in the engine is too low to sustain adequate lubrication of the moving parts. It is essential that the engine is not allowed to run with low oil pressure, or it will wear very rapidly. The oil pressure is maintained by a pump which pumps oil round the engine while it is running, and therefore there will be no oil pressure when the engine is not going and the light will come on. This may seem a little complicated to some readers, and just to confuse things a little more, an engine which is worn will have low oil pressure, and the oil warning light will come on when the car is at its normal running temperature. To put this another way, if a car is 'clapped-out', the oil warning light will come on when the engine is hot and running. This fact is fairly widely known, so it is not beyond the more unscrupulous car seller to disconnect this warning light so that it doesn't light up to warn you that the car is 'clapped-out'.

It doesn't matter if you don't quite understand all this; what matters is whether the oil light is working or not. The light should come on when you turn on the ignition, and go out as soon as the engine is started. If starting from cold it is OK on an older car if the light takes a couple of seconds to go out, but no longer. If it hasn't come on when you turn on the ignition, then it can't go out, so make sure this light comes on *before* the engine is started.

Before you start up, check that all the electrical equipment is working. *Wipers, horn, indicators, lights etc. should all work properly* **64** to **71**.

None of these things are serious in themselves and no car will manage several years of use without some repairs to these systems. Nevertheless, there is little point in buying a car which needs to have work carried out straight away, so if you do find that something is not working properly, ask the seller to attend to it before you buy, or ask for a price reduction.

The road test

At last we are ready to hit the road. Well, nearly ready anyway. There are just a couple of matters to attend to before we can get around to finding out whether the thing actually goes or not.

You've already checked the MOT, but remember that it is illegal to drive without tax and that it is even more serious to drive without insurance. So, make sure that you have some insurance cover. If you already have an insurance policy, then you will probably be covered to drive a car that doesn't belong to you, but normally this will only provide third party cover. This means that if you prang the car while you are testing it, you may be liable for the cost of putting matters to rights. If you don't have cover, then make certain that the seller has an insurance policy that covers you. Someone anxious to sell a car will often give a bland assurance that you are insured, when in fact you are not, If you did cause an accident – it's easy to do in a car with which you are not familiar – then you could find yourself with a bill which you'd be paying off for the rest of your life. If you are in any doubt, ask to see the insurance certificate.

The test drive is not simply a matter of a quick

drive around the block to see if it goes or not, but a series of tests which will show up any faults in the internal workings of the vehicle and hopefully prevent you from buying something which will be in need of expensive repairs in the near future. So, at last, this is it. Time to drive the car!

Each car is an individual, and has its own preferences about how it is started up. Ask the seller how much choke it likes and so on before you attempt to get things going. Turn the key and see what happens. Some cars are *difficult to start* **72** and this may only mean that some minor adjustment to the car's more sensitive parts is necessary. On the other hand, difficult starting can mean that the car is suffering from low compression problems, i.e. it is 'clapped-out'.

If the engine 'tries' to start, it sort of coughs but does not start and does this a few times before it starts running, then there may well be a problem with compression and it will need a major overhaul before very long.

You've already made certain that the ignition and *oil warning lights* **73** come on with *the ignition* **74**, so now we have to make certain that they go out when the engine starts up. As soon as the engine starts, watch the oil light and make sure that it goes out within a couple of seconds. If the engine is cold when you start up, then while you watch the oil light, it is worth *listening to the engine* **75**. A worn engine will make a rattling noise for the first couple of seconds, after which the noise will disappear. It takes a bit of dexterity to do both these tests at the same time, but it isn't too difficult.

Hopefully you will now be sitting in a most comfortable seat, with the engine quietly ticking over (having started first time) and no sign of life from the oil light. The 'ign' light may flicker a bit on tick-over: that is fine as long as it goes out as soon as the engine speed increases. Most engine problems show up when the engine warms up, so it's worth driving round for a few minutes just to get the feel of the car before you begin the more rigorous trials.

So far, during what has been a very thorough examination of the car, you have (on the whole) been using your eyes to look for faults. For the road test, your eyes will be rather occupied with the task of ensuring that you drive the car around without hitting anything and from now on you will have to pay a lot of attention to what you hear. Although it takes many years of experience before it is possible to say with any degree of certainty that a particular noise indicates the presence of a particular fault, you do not need this knowledge. All that is required is to listen and if you hear noises other than the smooth sounds of the engine purring away happily to itself, then you'll know that something is wrong. It doesn't really matter precisely what is at fault. If you hear horrible grinding and 'graunching' noises at anytime, then you can safely assume that there is something amiss and you will have to look elsewhere for your next car.

So, with the engine ticking over, *depress the clutch* **76**, but before you engage first gear, have a listen to see whether there is anything different about the noises the car is making with the clutch down. A squealing or grinding noise indicates that the thrust bearing is worn, and will need to be replaced in the near future. In some cars it is a big job as the engine or gearbox has to be removed to get at it. This is not a common fault however, so it is not very likely that you will encounter it.

Now *engage first gear* **77** and let the clutch in. It should start to engage an inch or so from the floor. Some variation is allowable, but beware of a clutch which starts to engage as soon as you begin to release it, or one which is at the end of its travel before it bites. It may only be a matter of adjustment, but it could mean that there is no adjustment left and the clutch is worn out. There will be a further clutch tests later on, so don't worry too much at this stage.

When you drive off, check that the car pulls away properly, and that the acceleration is all you would expect of the type of car you are testing. If *the general performance* **78** is sluggish, then the problem could be anything from a very minor fault such as a blocked air filter, or it could mean that the engine is about to expire due to old age. Later tests will reveal all!

Take the car up to fairly high engine revs in each gear before changing up, that is change up to the next gear a little later than you would do normally. If there is wear in the gearbox, then there may be a tendency for *jumping out of gear* **79** and it is most likely to do this when it under strain at high engine speeds. So, if you are accelerating along, and there is a sudden clang and you find yourself in neutral, costly repairs will be needed soon.

Note whether the all-round performance in each gear is satisfactory for the type of car. If the car seems to be under-performing, it may be that you are used to something with a little more guts, so don't expect too much from an economy class car. While you are checking the general behaviour of the vehicle, it is as well to include the *ability of the car to stop* **80** as well as its forward movement. Try pumping the brake pedal up and down a few times. This should not improve *the braking efficiency* **81**. If it does, then there is air in the hydraulic system which will have to be attended to.

While you are warming the car up, take a look at *the temperature gauge* **82**. This should rise to the normal position fairly quickly, then stay there.

In order to really put a car through its paces, you will need to find a hill, preferably about half a mile or so long. It doesn't need to be very steep, just something the car will roll down easily. Start at the bottom of the slope with a hill start. Make sure that *the handbrake* **83** holds the car properly without you having to pull on it excessively hard. Now pull away up the hill and see *how the car performs a hill start* **84**. It won't do quite as well as on the flat of course, but it shouldn't have too much difficulty.

When you get to the top of the hill, then it's time to drive the car down again. Rather surprisingly the downhill run is one of the most important tests for engine wear there is, so even if you live in one of the flatter areas of the country, it is worth a journey to a hill for this one.

Set off from the top of the hill and build up to about 30 mph in top gear. Leaving the car in gear, take your foot off the accelerator, and allow the car to roll down. Normally when you drive along the road, the engine is used to power the car and propel it along. In this test we are doing the opposite and using the car rolling down the hill to turn the engine. If the engine is worn, oil will get into the combustion chambers, which during normal running would be burnt up. In this instance, any oil which gets into the combustion chambers is not burnt up as you don't have your foot on the accelerator, so it accumulates in the combustion chamber. It doesn't matter if you don't understand all this: it's only mentioned for the benefit of readers who may be interested in the reasoning behind this strange ritual.

While you are running down the hill, it is a good

opportunity to listen out for any rumbling, knocking, 'graunching' or grinding *noises* **85**. The engine will be fairly quiet relative to the roadspeed at this point, so it will be easier to hear if there is anything untoward. In addition, there are some faults which can only be heard on the 'overrun', that is when the engine is being turned by the momentum of the car, so make a note of anything you hear. It is difficult to be specific about exactly what to listen for, but noises which don't sound right and only appear under certain conditions are a bad sign.

When you get near to the bottom of the hill, it's time to test for *engine wear* **86**. If you have a friend with you, then ask him or her to look out of the rear window for you. If you are on your own, then you will have to manage with the rear view mirror. If you have to stop at the bottom of the hill that's fine, put the gears into neutral and give the engine a really good rev up. If you don't have to stop, put the car into third gear and accelerate away as fast as possible. The car should of course accelerate away smartly, but we're not too bothered about that here. What we are concerned about is how much oil got into the combustion chambers while we were cruising down the hill listening for funny noises. When the engine is revved up, this accumulated oil will burn off and give off characteristic blue smoke in the process. A puff of smoke here is inevitable, but if the car behind you suddenly disappears in a blue smoke-screen which apparently appeared from nowhere, then you are at the wheel of a car with a badly worn engine which will need extensive repair work before too long.

If all went well on the hill, then the next set of tests will be carried out on a long clear stretch of road, preferably one where it is permissible to drive at 70 mph and where there is not too much traffic about.

Drive along in third gear, not too fast, then take your foot off the accelerator. As the car slows down, accelerate away, then repeat the process a few times it makes the ride rather reminiscent of early attempts at driving as you jerk along, but don't worry about that. Use your ears to listen for any rumbles or 'graunches' which occur during acceleration or deceleration, and in particular, listen out for any clonking noises as you change from acceleration to deceleration. Clonks indicate some *wear in the transmission system* **87** and these are always expensive to put right, particularly on front wheel drive vehicles.

Next drive along in a straight line at around 40 mph and take both hands off the steering wheel. Do not take them right away, just hold your hands an inch or so away from the wheel and see what happens, the car should continue along in a straight line. On a motorway, it should even go round bends without assistance due to the camber or slope built into the road surface. However, it is best not to depend too much on that. If you are on a straight stretch of road and the car seems to want to leave the road by *pulling to one side* **88** or the other, firstly prevent it from doing so. Secondly, make a note of the fact that there is a fault somewhere in the steering system. This is a tricky one as there are many possible causes which can result in the car pulling to one side. It may be caused simply by one of the tyres being rather low on air pressure, or, at the other extreme, the car may be twisted, as a result of a bad accident. An expert may be able to tell you what the problem is, but as usual, if it is not just a slightly flat tyre that is causing the problem, then err on the side of caution rather than buying a car which may be dangerous.

Start to increase speed from 30 to 70 mph. Don't rush: we're not interested in how good the acceleration is at this point. Increase speed slowly until you are going at a steady 70, then slowly decrease speed again. The car should travel smoothly at all roadspeeds, and you should not notice anything untoward. You may find however that at certain roadspeeds there will be some *vibration* **89**, or *the steering wheel develops a wobble* **90** which disappears as the speed is increased or decreased. Such problems are caused by wear or imbalance in the transmission system. Again, it could be a very simple problem such as a wheel that needs to be balanced – cost about £2 – or something pretty horrendous which could cost a small fortune to put right. Unfortunately only an expert can tell, so we will just have to get some further advice: free advice of course!

Here's where the road test gets to be really exhilarating as we have to *test the brakes* **91** properly! Brakes can work fine when you are pottering about, but you need to know how they will perform when you really need them. 30 to 40mph is the sort of speed you should use for this test: any faster and it could be dangerous. Before you do it though, there are two things you should do. The first is pretty obvious. Ensure that there is nothing behind you and it is best to ensure that there is nothing coming the other way either – you don't yet know what the car is going to do under severe braking. Secondly, tell yourself before you start that if anything goes wrong, you will *release* the brakes. (There is often an instinctive reaction to brake harder under unusual circumstances.)

All ready? Try an emergency stop. If all is well then you will rapidly decelerate in a straight line. If however one of the brakes is not working properly, the vehicle will slew over to one side due to the uneven braking. This can be rather exciting, but if the car does begin to deviate from the straight and narrow, release the brakes immediately and straighten up. Although it could be argued that the public highway is not the place for testing the effectiveness or otherwise of a vehicle's braking abilities, it is far better that any faults are discovered when there is nothing around to bump into.

This type of problem in a braking system is not uncommon and does not mean that the car as a whole is dodgy It simply means that it will have to spend an hour or two in a garage being put right and somebody will have to pay for this.

Now we'll try out *the cornering* **92**. There's no need to burn any rubber here, we'll assume that the design of the car will ensure that it goes round corners without leaving the road. What we need to know is whether all is working as it should. So try going round a corner in a normal manner, then try another by approaching slowly, then accelerating round. Listen while you do this. Clonks, grating noises, etc. are as usual bad news and indicate that something is wrong somewhere.

Notice too how the car feels as you test *the cornering ability* **93**. If it feels really peculiar like you're in a boat or hovercraft, then it could be because you're driving one of the smaller Citroens or Renaults and are unused to the way they handle. On the other hand, a twisted car can feel rather strange on a bendy bit of road, so you could be driving a car which has had an accident at some time.

The engine should be nice and hot by now – look at the temperature gauge to make sure that it isn't overheating – so we are ready for the next test. Stop the car and apply the handbrake. Now engage second gear, and attempt to pull away with the brake on. If the car starts to move pull the brake on harder. Give the engine plenty of revs, as if you are attempting a very steep hill start. Although the engine will be a bit noisy, listen out for a knocking *noise* **94**. Big end knock (a problem which is even worse than it sounds) will occur when the engine is under load. Keep on trying to pull away without stalling the engine by keeping the revs up and slipping the clutch **95**. If you find that the clutch can be fully released without the engine stalling, then it is well on the way to being worn out, or oil is leaking from the engine onto the clutch. Either way, some expensive repairs will be imminent so the car doesn't look like a good buy.

The last thing to check whilst the engine is hot is *the oil pressure* **96**. Let it tick over slowly. The oil light, unlike the ignition light should not flicker at all. If it does, then the engine doesn't have much life left. The seller may tell you that the fault lies in the device which senses the oil pressure. This may be the case as these 'sender units' can be unreliable. However, they are cheap to replace and if this is where the problem lies, then he won't mind fitting a new one to prove his point! Never buy a car if you see the oil light flickering.

By now you will probably know a lot more about the car than the person who is selling it. With any luck it will have come through the road test without any major faults being revealed, so it's time to take it back to base for a few final checks.

There is one more test worth doing on the way back, which ideally involves driving the car at high speed over a newly ploughed field. Unfortunately this can sometimes upset the owner of the field and it doesn't tend to go down too well with the owner of the car either. All is not lost however and the next best thing is usually available in the form of a road full of potholes that the local authority hasn't got round to fixing yet. Drive along at a brisk pace and listen out for any rattles or clonks which could indicate that there is a problem with the steering or suspension systems 97.

A few final checks

If there is anything seriously wrong with the car, you will almost certainly know about it by now, but there are a few things which are worth looking at while the engine is warm.

When you get back to the seller's premises, leave the engine running and open the bonnet again. You need to be very careful at this point, as not only do you face the danger of parts of the engine being rather hot, but there is a fan going round at high speed behind the radiator. People have had serious damage done to their hands by cooling fans, so keep your fingers well clear. Ties, loose clothing and long hair should also be kept out of reach of this rather dangerous device.

Remove *the oil filler cap* **98** – the one on the top of the engine which was examined at the start of the test. Some oil will probably splash out, but don't worry. Get your friend or the seller to rev the engine a little. If more oil splashes out that's a good sign, but you don't want to see any smoke coming out. It probably won't unless you laid a smoke screen during the road test, in which case you will have decided not to buy the car anyway, but it's worth checking.

If you know something about engines, have a listen to the front end for *a rattling noise* **99** which could indicate a worn timing chain. If the internal organs of the motorcar are beyond your understanding, then don't worry too much. You've listened out for all the serious noises and listening for noises under the bonnet takes a little experience due to all the other sounds made by all those bits and pieces moving around.

Stop the engine, and have a general look round at the contents of the bonnet cavity. You will see that there are *tubes and pipes* **100** going from place to place in apparent disarray, emerging from the most unlikely places, all with a purpose of their own. The thicker ones usually carry water, while others are used to convey petrol, hydraulic fuid, air and so on. Most of these tubes are made from rubber and will not normally last the full life of the vehicle. This means that some of them will need to be replaced at some time and whereas this should not in itself put you off buying a car, you don't want to be spending money on repairs straight away.

Examine all the pipes for signs of leaks or cracks in the surface. If they are starting to perish you may be able to negotiate a price reduction.

The same applies to *the radiator*, **101** – that rather hot thing at the front of the car. If you happen to be looking at a car with an air-cooled engine, then you are probably looking at an early model produced by Citroen or Volkswagen which do not have one. Look for signs of water leaking out. If there are any leaks, you will probably have noticed some steam escaping when you raised the bonnet after the road test. Look for a rust coloured stain, particularly near where the rubber pipes join on, which will indicate the presence of a leak. A leaking radiator isn't a disaster, just something which will need to be replaced. Remember that the filler cap must not be removed when the engine is hot.

What next?

What next indeed! It's all very well doing all these fancy tests on a car and noting everything down on the test sheets, but what do we make of it all? What about the funny noise the car made going round a corner? Is it serious? Does it matter? How do I find out? The answers to these questions can be found in a variety of places and as I said earlier, you can get expert help which doesn't cost you anything. All you'll need is a bit of time and the co-operation of the seller.

The situation differs a little depending on whether you are involved with a private sale or a motor trader. We'll deal with the latter first.

If you've discovered a few minor faults like the windscreen wipers don't work properly, one of the doors doesn't shut and there is a hole in the exhaust system, then ask him to attend to them. If you are buying from a dealer you should expect any faults however minor to be put right before you buy.

If you've discovered something you're not sure about like a problem with the steering, or crunching noises that you cannot identify, then simply ask the seller what it is. He may tell you it's just a loose phlange nut on the prop shaft universal coupling and nothing to worry about. This may be true, or it may be something serious: you have no way of knowing. Fortunately it doesn't matter. If you are otherwise happy with the car, tell him you'll buy it when he's fixed it. Don't be dissuaded from this position and don't pay a deposit either. Ask him when it will be fixed so you can call back to re-test it. When it has been attended to, repeat the test which first revealed the fault. If the noise has stopped, then you can reasonably assume that whatever the problem was which caused the symptoms has now been fixed and the precise mechanical details need not concern you.

When buying from a motor trader you should always insist you receive a full years MOT, that way you can be reasonably certain as far as safety is concerned, all is in good order. This is particularly important with regard to steering which is not always easy to check using the tests in this book. (See the section at the end of the book, "How to use the test sheets".)

Human nature being what it is, it is possible that a car dealer carrying out an MOT test on a car that he is about to sell will be just a little more lenient that he would otherwise be. It is also possible in some cases that he would pass a car with a serious defect. However, by doing so he'd risk losing his tester's licence, and it's not a risk many people would be prepared to take. There isn't a lot we can do about this one, although it won't cost anything to have the steering re-checked by an independent engineer if you really want to be certain.

There's one crafty dodge which is really worth knowing about and you should always use it when buying a car from a trader. If all seems fine and you are happy with the car and think it's a good buy, tell him you are interested, but you'd like to think about it for a few hours, discuss it with your wife/husband/bank manager etc. Ask him to keep it for you until the next day when you'll let him know definitely either way. Most people want to have a little time to think over such a big decision, so he will almost certainly oblige.

You will remember that the dealer does not have to record his ownership of the vehicle on the registration document. At the top of your test sheet you therefore wrote down the name and address of the previous owner, i.e. the person who sold it to the garage. This person will know more about the car you have been looking at than anyone else. He'll know what its fuel consumption was like, what problems it has had and quite a few other things you won't find out by testing it. This person will have traded the car in, and bought another from the dealer. He has therefore been paid for it, no longer owns it and has no further interest in it. Most people are on the phone these days and you'll probably find the number in the telephone directory. If not, it's worth a trip to call round in the evening.

Apologise for troubling him and explain that you are thinking of buying the car he recently sold to the trader. Ask if there was anything wrong with it when it was sold. You do have to be very tactful here. You may be asking him to admit that he off-loaded a duff car onto his local garage. You could make it sound like you're not too sure whether you can trust the garage proprietor and want to check up on him. He doesn't have anything to lose by telling you and in my experience people are very helpful in this situation. They will gladly tell you if they've had problems with the car, what the fuel consumption was like, what a nice time they had when they went to Spain in it, and how well it behaved going over the Alps, how much granny liked being taken out and so on. Given the chance, most folk seem to enjoy being helpful. I know someone who tried this recently. The previous owner took great delight in telling him what an awful car it was, how much trouble he'd had and advised my friend not to touch it with a bargepole!

Do ask what the mileage was when it was sold to the garage. If the clock has mysteriously lost a few thousand miles, then you should report the matter to the local Trading Standards Office and go elsewhere to look for a car.

The situation is rather different in the case of a private sale, as you will be dealing directly with the last owner and you cannot expect that he will be as forthright about any problems he's been having. If the vehicle seems to be basically sound, but there is a funny noise you're not too happy with, or the steering doesn't feel quite right, tell him that you are very interested, but you would like to have an expert second opinion.

There are two sources of advice available which need not cost you anything. The first is applicable to steering, brakes and so on; the safety aspects of a car, and the second is more general.

In most towns of any size there are garages – usually the type which specialize in exhausts, tyres etc – which display a sign which says 'Free Safety Checks'. They don't say it has to be your own car, in fact they don't make any stipulations at all. In most cases you don't even need an appointment. You just roll up and ask for a free safety check. They are usually concerned with the tyres, steering and brakes. Since the steering was difficult to evaluate properly on the test due to the wide ranging differences between cars and the difficulty of devising a simple test applicable to all types, then a safety check is very worthwhile.

It doesn't take very long, 5–10 minutes. An expert can very quickly establish what the problem is and not only will they tell you what's wrong, they'll tell you what it will cost to put matters right.

Even if the car you are looking at doesn't seem to have anything wrong with it, you could have a safety test done just to be on the safe side – it's free, so why not. The seller probably won't mind as you are unlikely to go to this amount of trouble unless you are seriously interested in buying the car.

Some people may feel that they are being just a little bit cheeky by taking advantage of this expert advice free of charge. However, there is no harm in doing so. The people who offer this service do so in the hope that they will find something wrong and that you will ask them to carry out the repairs. You may well be able to negotiate a price reduction to cover any work which needs to be done and you may also decide to give the work to the garage which detected the problem. You may also consider that the safety check is likely to be very thorough; they have a vested interest in finding as much as possible wrong with the car. No leniency here!

The other source of help you can turn to is your

good old-fashioned friendly neighbourhood garage. Although such establishments were at one time facing extinction in the face of competition from the large 'main dealer' outfits, there does seem to be a move back towards these traders. If you already possess a car, then you probably have work done from time to time at a garage where you are known and this can be an advantage, if you don't have an established relationship with a local repair workshop, then this may be the time to start. Ask around the car owners you know to see which garages have a reputation for being helpful and providing a good service.

It is common practice for people to turn up at a repair workshop to get an estimate for work that needs to be done. You can phone in advance if you like, but usually you don't need to bother. Just explain the situation – you don't have to tell them you haven't bought the car yet if you don't want to, but tell them that the car has started to make a funny clonking noise / doesn't go properly / has steering pulling to one side etc. In many cases an experienced mechanic will be able to tell you in a couple of minutes what the problem is and what it will cost to put right.

You may not get service like this from one of the big chain dealers, but you'll be surprised how helpful some places can be.

To buy or not to buy?

Making decisions is something which human beings seem to find difficult on a good day. It's a problem which has afflicted people of all races and cultures throughout the ages. Particularly difficult are those decisions which involve having to part with large amounts of money.

You have just spent an hour or so going over, crawling under, poking and prying, testing and checking a car to find out if there is anything wrong with it, and as if that wasn't enough, now you have to make a decision about whether or not to buy it. It isn't easy, but at least you have a lot of information to help you. All you have to do now is to try and make some sense of it. There are several possible results from your examination procedure, so we'll have a look at each possibility in turn.

The first one is that the car you are looking at has obviously been well looked after, has been serviced regularly from new, only had one owner and has a genuine low mileage and you can't find a single thing wrong with it. A find like this occurs rather less frequently than a blue moon and you aren't very likely to find another, so the decision to go ahead and part with your money shouldn't be too difficult.

The second possibility is that you found a car which at first sight looked OK, but as you delved into its secrets, you discovered that it had holes in the chassis, made peculiar noises, had not been cared for very well and would probably expire in the near future. Cars like this are unfortunately not quite as unusual as the previous type, in fact they are all too common and you will probably come across a few in your quest for the perfect motorcar. The decision about whether or not to buy is again not difficult. The time you have used to give the car a thorough examination is not wasted, but time well spent. You have probably saved yourself a great deal of money.

The third, and by far the most likely possibility is that you have given the car a very thorough examination and there are one or two points which give you a little cause for concern. You are most unlikely to find a car with no faults and what you have to decide now is whether or not the faults you have found are sufficient to put you off buying.

Although decisions are not easy, this one isn't as bad as it first seems. Apart from a car which has corroded away underneath, there isn't really very much that can go wrong that can't be put right. What matters is how much it will cost to be put right. If you have had the car looked at by an expert either at a free safety check or the local garage, then you will already have some idea of the costs involved. If you have been able to identify the problems yourself and you have for example discovered a worn shock absorber, or a clutch which needs replacing, then it is simply a matter of telephoning a garage to ask for an estimate for the work. In many cases, estimates for standard jobs can be given over the telephone, although for some repairs, they will have to see the car first.

Having established what it will cost to put matters right – don't forget that an estimate does not usually include the dreaded VAT, so you will need to add this on – then you can haggle with the seller. He may decide to sell the car to you at a lower price in view of the work which needs doing, or he may decide to hang onto the vehicle in the hope of selling it to someone less vigilant than yourself.

Deciding whether or not to buy a car is never easy, and by some strange irony it can be even more difficult when you know as much as you do about the vehicle. At the end of the day, it depends on what you want, how much you have to spend, and what you are prepared to put up with. Never be in too much of a hurry and if you are in any doubt, don't buy. There are plenty of used cars for sale. A large number of them are heaps, but there are plenty of good ones around too. You just have to sort through till you find one.

Even after all this prying and probing, you may have found a car which seems fine, but you don't feel absolutely certain about it. You may have doubts about your own ability to carry out a thorough test, although by now you will probably be fairly competent. Or you may be buying a car in a higher price range and you are not happy about investing so much money without a full expert examination. You may have missed something and as we have said before, it is beyond the scope of a general work of this type to cover every single possibility in every type of car. All we can do here is be almost certain!

In this case you may decide to employ the services of a professional vehicle tester. The AA and RAC both provide this service, and you will find people in the Yellow Pages. It isn't cheap – nearly £50 + VAT is the going rate at the time of writing, but you may feel it's worth it for the peace of mind. You don't need to pay this much however. Many garages will give a car a good going over for a fraction of this amount. It's not a service they tend to advertise, but most are happy to do it.

If you have been thorough, you will probably inspect a few cars before you find one that you are happy with. If you'd had your 'rejects' professionally examined you'd have wasted a lot of money. A professional would have been more thorough of course, and may have picked up a few things you missed, but on the whole you will probably do nearly as well yourself.

Once you decide to go ahead and buy the car, then before you part with any money, get the seller to confirm in writing that he is the owner of the car, or that he is entitled to sell it, that it is not the subject of an outstanding hire purchase agreement and that to the best of his knowledge there are no undisclosed faults and the mileage on the clock is genuine. You will also be well advised to obtain a receipt for the money. A standard form is included in Appendix II, which covers all these points, and you can use this if you wish.

Conclusion

Unfortunately a large proportion of used cars on the market are in a pretty awful state. Some are simply 'clapped out', whilst others are downright dangerous. The chances of finding a good one the

first time you look are not high and you'll probably look at a few before you find one you are happy with. It will be time well spent – a car is a costly item, and with the cost of garage bills soaring ever higher, it's worth getting it right and not being sorry later.

In the process of examining a few cars you'll develop an expertise of your own, the first few will take a little while but once you get the hang of things you'll do a quick once over in a few minutes deciding in some cases that a full examination is not worth the time and effort.

Perhaps one of the most important things you'll gain is the satisfaction of doing something for yourself, not having to rely on other people to do things for you, and not, as many people do, relying on luck either. You will be in charge of the situation and making your own decisions. Good hunting!

Appendix I: How to use the test sheets

Each test on the sheet has a number beside it which corresponds to the section in the test which explains how to do it. So if you can't remember what to do while you re filling in the test sheet, then refer back to the book.

To the right of the list of tests are two columns, two of which are for filling in, one for your comments, and the other simply headed pass/fail.

In the pass/fail column you simply put a tick or a cross depending on whether the car came through the test with flying colours or not. This seems simple enough, but life isn't really like that in practice. There are some tests which we carry out which will reveal a fault or indicate that the car is fine on that particular score. If for instance the vehicle makes a horrible clonking noise when you accelerate round a corner, then it's failed. If it doesn't then it passed. In many instances however, it's a matter of degree. An acceptable amount of rust under the wheel arches is a matter of judgement, and how much you intend to spend on your next car.

The comments column is there for just this type of contingency. You may feel that you are not really qualified to make comments on something you know nothing about. Don't worry, politicians do it all the time, and are undeterred by the fact that they don't usually know anything about what they are spouting on about. In any case, you will not need to use this column for every test – if you did it would take far too long to evaluate the results. All that is required is an occasional observation along the lines of good, fine, poor, awful, where there is no obvious pass or failure.

Of course judgement varies from one person to another. What one person would consider to be a smart interior, another would regard as being pretty tatty. That's fine. This problem crops up in most spheres of human activity – it's the subjective factor. Since it's you that will have to live with the car if you buy it, then your opinion matters most. If you don't like the look of things, write an appropriate comment.

By the name of each test, there may appear an asterisk, two asterisks or an E. These are 'dire warnings'! If you note a failure or an adverse comment on a test which has an asterisk, then this indicates that you have found a potentially serious fault, the sort of thing which may need some very expensive attention in the foreseeable future. It is there as a warning to be careful. A twisted appearance to the car does not mean necessarily that the car is bent beyond redemption. Nevertheless, it is always better to err on the safe side and not buy something which could be an expensive mistake. If you are not happy with something about the vehicle which has two asterisks marked against it, then you are looking at something very dodgy indeed and would do best to look elsewhere.

An E as a dire warning indicates that you will need expert advice as to whether the problem is serious or not. Most of these will be covered in a free safety check and if you are in any way unsure, then have a safety test carried out.

The absence of an E or an asterisk does not imply

that the problem is not serious, just that it wasn't considered to be quite serious enough to warrant a dire warning. If for example, the car fails on the emergency brake, then work will have to be done to put it right. However, faults of this type often crop up during the lifetime of a car and as long as the seller will drop the price to cover the work, then the car may still be worth buying.

You will notice that some items in the pass/fail column are marked with an 'M'; this indicates that this item forms part of the MOT test. If the vehicle has recently passed an MOT, then you should be able to assume that all is in order. However, if you want to be really certain, carry out the test yourself.

The last test **102** on the sheet may be a bit of a surprise after all you've done up to now, but it is one of the most important. It would be very easy to become so embroiled in the quest to find a perfect motor car that this small point was forgotten. But after all this rigorous investigation, you must consider whether you actually like the car or not.

You may find a car with no faults at all, but you may simply dislike it. There's little point in buying something you don't like just because it looks like being a bargain. You might be better off with something which has a few minor faults, but which doesn't affect your street credibility too much and with which you feel happy.

In the end it's all up to you and you'll have to make your own decision based on what you have found out about the car. The one thing you can be sure of is that you will stand far less chance of being ripped off than you would if you had been content to buy the car without a proper inspection.

Appendix 2: Vehicle test sheet

Particulars of vehicle registration no ________________

Make ________ Model ________ Year ________ Engine size ________

Test no.	Description of test	Comments	Pass/ Fail
1	General impression of vehicle		
2	Reason for sale		
3	Any faults declared by seller		
4	Name and address of seller Name and address of previous owner/registered keeper (from reg. document)		
5	Number of previous owners		
6	Date of last sale		
7	Body/chassis no (VIN)		
8	Engine no		
9	Record of insurance write off**		
10	Mileage on MOT certificate		

Appendix 2. Vehicle test sheet

Test no.	Description of test	Comments	Pass/ Fail
11	Vehicle recorded mileage		
12	MOT expiry date		
13	General impression under bonnet		
14	Serious oil leaks		
15	Engine no. same as reg document		
16	Body no. same as on reg document**		
17	Service record available		
18	Oil level*		
19	Oil condition		
20	Oil filler cap – white deposit*		
21	Radiator coolant level		
22	Oil in radiator water		
23	Antifreeze (winter months)		
24	Condition of oil filter		
25	Battery fluid level		
26	Condition of tyres		M
27	Uneven tyre wear		
28	Respray		
29	Bumpers parallel to the ground*		
30	Twisted appearance*		
31	Security numbers*		
32	Doors close/fit properly*		
33	Boot and bonnet fit properly*		

Test no.	Description of test	Comments	Pass/ Fail
34	Distortions in bodywork*		
35	Rust holes in sills		M
36	Repairs in sills		M
37	Repairs/rust on doors		
38	Bodywork general condition*		
39	Rusting under bonnet*		M
40	Jacking points		M
41	Condition of wheel arches		
42	Wheel play: rear E		M
43	Wheel play: front		M
44	Bounce test		M
45	Condition of underseal		
46	General look for rust*		
47	Rust in box sections**		M
48	Evidence of repairs*		M
49	Join across width **		
50	Rusting of brake pipes		M
51	Rusting of petrol pipes		
52	Condition of petrol tank		
53	Condition of exhaust system		M
54 55	Oil leaks! engine/ gearbox* Leaks at back of wheels		M
56	Condition of tyres		M
57	General condition of interior		

Appendix 2. Vehicle test sheet

Test no.	Description of test	Comments	Pass/ Fail
58	Rust in floor		
59	Damp floor		
60	Seam across floor**		
61	Comfort of driver's seat		
62	Ignition light working		
63	Oil light working		
64	Wipers		M
65	Horn		
66	Lights. Dip and main beam		M
67	Screen washers		M
68	Heater motor		
69	Brake lights		M
70	Rear screen heater		
71	Other electrical equipment		
72	Ease of starting		
73	Oil light goes out		
74	Ign. light goes out		
75	Noises when engine starts up*		
76	Clutch thrust bearing*		
77	Clutch adjustment		
78	General performance		
79	Jumping out of gear		
80	Brakes. General performance		M
81	Brakes. Better after pumping		M
82	Temperature gauge		

Test no.	Description of test	Comments	Pass/ Fail
83	Hill start. Handbrake		M
84	Hill start. Performance		
85	Unusual noises. (down-hill)*E		
86	Smokescreen**		
87	Wear in transmission system*E		M
88	Pulling to one side E		M
89	Vibration at certain road-speeds E		
90	Steering wobble at certain speeds E		M
91	Emergency brake		M
92	Noises when cornering E		M
93	Roadholding on corners E		M
94	Engine noise under load*E		
95	Clutch wear*		
96	Oil light comes on or flickers**		
97	Noises on bumpy road E		
98	Smoke from oil filler hole*		
99	Rattling noises from engine		
100	Condition of rubber pipes		
101	Radiator leaks		
102	Do you like the car?		

Appendix 3: Certificate of ownership.

I ______________________________ (Name)

Of ____________________________________

____

______________________________ Address,

Confirm that I am the owner of the vehicle:

Make __________ Model __________ Reg no __________

Or that I am legally entitled to sell it on behalf of the owner, and that it is not the subject of a credit agreement.

I acknowledge receipt of the sum of £ ____________ in full payment for the vehicle.

I confirm that to the best of my knowledge:

The vehicle does not have any faults which I have not disclosed, and that the recorded mileage of

__________ miles is correct

Signed ______________________ Date ______ ____